OPTIMIZATION AND BUSINESS IMPROVEMENT STUDIES IN UPSTREAM OIL AND GAS INDUSTRY

WILEY SERIES ON OIL AND GAS TECHNOLOGY

R. Winston Revie, Series Editor

OPTIMIZATION AND BUSINESS IMPROVEMENT STUDIES IN UPSTREAM OIL AND GAS INDUSTRY

SANJIB CHOWDHURY

Library of Congress Cataloging-in-Publication Data

Names: Chowdhury, Sanjib, author.
Title: Optimization and business improvement studies in upstream oil and gas industry / Sanjib Chowdhury.
Description: Hoboken, New Jersey : John Wiley & Sons, Inc., 2016. | Series: Wiley series on oil and gas technology
Identifiers: LCCN 2016015712 (print) | LCCN 2016027748 (ebook) | ISBN 9781119100034 (hardback) | ISBN 9781119246589 (pdf) | ISBN 9781119246572 (epub)
Subjects: LCSH: Petroleum engineering. | Gas engineering. | Petroleum–Prospecting. | Natural gas–Prospecting. | Petroleum industry and trade–Management. | Gas industry–Management. | BISAC: TECHNOLOGY & ENGINEERING / Power Resources / Fossil Fuels.
Classification: LCC TN870 .C5185 2016 (print) | LCC TN870 (ebook) | DDC 622/.3380684–dc23
LC record available at https://lccn.loc.gov/2016015712

Set in 10/12pt Times by SPi Global, Pondicherry, India

Printed in the United States of America

10   9   8   7   6   5   4   3   2   1

In memory of my beloved parents and elder brothers,
whose blessings I count on and keep me moving.

CONTENTS

PREFACE

Oil and gas exploration and production (E&P) is a complex process involving a series of activities that are costly, risky, and technology intensive and require specialized manpower.

With a rise in the global oil demand and fast depletion of easy reserves, the search for oil is now directed to more difficult areas, such as deepwater and arctic region, and future production is expected to come from increasingly difficult reserves (deeper horizon, heavy oil, etc.). As a result, E&P activities are now even more costly, risky, and dependent on cutting-edge technology. Therefore, it is necessary to optimize resources, opportunities, and costs and improve business performance in the entire gamut of E&P activities.

Furthermore, due to inherent uncertainties involved in E&P activities, where input is deterministic but output is probabilistic, it is imperative that oil companies use their capital and resources judiciously. In this regard, various optimization and business process improvement techniques have been discussed in this book, which have the potential of saving substantial amounts of money and improve organizational efficiency.

The book is unique as it delves into the core and functional areas covering a wide range of operations and processes in the upstream oil and gas industry. The book contains 11 real-life studies conceived and developed by the author covering an extensive array of activities and processes, namely, optimization of strategies, resources, and costs; improvement in business performance and operational productivity in core activities; identification and removal of inefficiencies in core and functional areas; standardization of consumption of materials; development of uniform standards for offshore safety alarm system; simplification of business processes for enhancing organizational efficiency; and improvement in human resource productivity. The book is replete with examples of quantitative and qualitative techniques, such as linear programming, queuing theory, critical path analysis, economic analysis, best practice benchmark, and business process simplification. The quantitative techniques are based on mathematical treatise that has universal applicability and acceptability.

The book will be of immense interest to practicing managers, professionals, and employees at all levels and disciplines in oil and gas companies, especially upstream companies to improve their respective operational areas, business processes, systems, and organizational efficiency. It will also be useful to academicians; scholars; educational institutes; energy research institutes, especially dealing with oil and gas; consultants; and others, as the work is rich and replete with the application of quantitative techniques in real-life problems. The work may also be used as a practical guide by professionals to enhance organizational efficiency and performance and optimize resources. The availability of such type of book has not yet been reported in the market or in the industry. Practicing managers and professionals at all levels in oil and gas industry would find it useful to know the nuances of optimization and emulate these studies in their respective domain of work, thus immensely benefiting their organizations. All these studies have the potential of saving substantial amounts of money, besides improving organizational efficiency. One can ignore the book, if one can afford to splurge substantial amounts of money!

Review exercises have been included at the end of each chapter, which will help readers to recapitulate the key learnings and test their understanding about the subject. It will also assist the readers to focus on important areas and revisit the relevant portions to capture their essence. Review exercises are mostly related to topics discussed in the respective chapters. But a couple of questions in Chapters 2, 3, 4, and 6 are a bit more technical and specialized in nature, which are marked as "Advance." These have not necessarily been discussed in detail in these chapters and require further studies. These are meant for subject matter experts (SMEs) and E&P professionals dealing with these jobs.

An enriching list of "References, Useful Links, and Further Reading" have also been incorporated at the end of each chapter, which will help the interested readers to learn more and go deep into the respective subjects as per their need.

The views expressed in the book are those of author's and do not necessarily reflect the views of any organization or entity or individual.

ABOUT THE AUTHOR

Dr. Sanjib Chowdhury is working in a major oil company and has over 30 years of experience in cross-functional areas of strategic and corporate planning, optimization and business improvement, HR, and general management in upstream sector. He has held responsible positions and carried out many impressive work and studies in these areas—this book is a testimony to that. He holds B. Tech., M. Tech., and Ph. D. degrees in Industrial Engineering and Management all from the Indian Institute of Technology, Kharagpur, and published several technical papers in journals of repute. He mentored several MBA students and junior professionals and was a guest faculty to business schools.

Esteemed readers are invited for critiquing and suggestions for the improvement of this book (optimize.upstream@gmail.com). The author wishes to share more such actionable ideas for enhancing revenue, saving cost, and improving productivity, and so on, worth substantial amounts of money in the future.

ACKNOWLEDGMENTS

I am grateful to almighty for bestowing me with the privilege of working at various geographical and cross-functional areas in the upstream oil and gas industry, which has enriched my professional knowledge and experience. This book relies hugely on this experience. The studies presented in this tome were conceived and developed by me. It seemed initially a daunting task, but on the whole, it was greatly educative and satisfying.

I am thankful to all those who had directly or indirectly helped me in completing these studies by providing information and data, helping me to understand different processes and perspectives, and engaging with me in professional discussions. I am especially thankful to Prof. Pradip K. Roy of Indian Institute of Technology, Kharagpur, for his encouragement, support, and advice; A. Pandey for extensive review, critiques, and suggestions; and A. Dutta and S. De for reviewing certain chapters and providing useful suggestions. All these have improved the quality of the book. I am grateful to my family for supporting me in this endeavor.

I will be immensely happy if readers find the book useful and apply the concept and learning in their work, organization, or business.

Finally, I live on blessings of my late parents and elders, which give me strength and keep me moving. I dedicate this small work in their memory.

DR. SANJIB CHOWDHURY

1

OPTIMIZATION AND BUSINESS IMPROVEMENT STUDIES IN UPSTREAM OIL AND GAS INDUSTRY: AN OVERVIEW

1.1 INTRODUCTION

1.1.1 Importance of Oil and Gas

Oil is inarguably the most important economic commodity and source of energy in today's world. It has shaped contemporary civilization and is intricately interwoven with our daily life touching every household. It fuels world economy, propels industrial growth, and impacts on nation's well-being. Oil accounts for one-third of world's energy need, while oil and gas together meet more than half of global energy demand. It will continue to dominate global energy mix in the foreseeable future. Oil is not only an economic commodity, but it has great strategic value too. The geopolitics of oil is well known, and the world has seen fierce disputes, even wars among nations for oil and gas. It influences world economy to such an extent that no country however mighty or humble can ignore it.

Life is unthinkable without oil and gas in the present-day world, which has pervaded not only our daily life but also deeply entrenched in the nation's economy encompassing all sectors including domestic, industrial, agricultural, transport, and other segments. For example, essential products, such as petrol, diesel, domestic gas, kerosene, naphtha, fuel oil, fuel gas, lubricants, wax, and so on, are derived from crude oil. It is a major component in many important products, such as fertilizers, organic chemicals, industrial chemicals, drugs, detergents, insecticides, cosmetics, and so on. It is also used in manufacturing household containers, furnitures, building materials, synthetic rubber, plastic goods, nylon clothes, CDs, DVDs, and many others. The transport sector is heavily dependent on it, and the world will come to a grinding halt without oil and gas. Ships, airplanes, trains, buses, cars, and so on, will stop plying; machineries, farm tractors, and factories will stop running; and industries using oil/gas as feedstock will close down [1].

Optimization and Business Improvement Studies in Upstream Oil and Gas Industry, First Edition.
Sanjib Chowdhury.
© 2016 John Wiley & Sons, Inc. Published 2016 by John Wiley & Sons, Inc.

1.1.2 Early Use of Oil and Gas

There are many evidences and stories connected with the use of petroleum, especially oil and bitumen, in the ancient times. The "eternal fires of Baku" were the result of the ignition of oil and natural gas from seepage, the "tower of Babel" was constructed using bitumen as mortar, the basket in which baby Moses was hidden was believed to be made waterproof using bitumen, and Persians set alight the streets with sprinkling oil when Alexander the Great visited Persia. The multiple evidences suggest that in earlier days oil was used in Egypt, Persia, and Mesopotamia for heating, lighting, and paving roads. The records also suggest that North American Indians used petroleum as medicine, Mexican Indians valued bitumen as chewing gum, and Chinese were believed to be drilling wells using bamboo canes. Many famous explorers mentioned about it, for example, Sir Walter Raleigh wrote about it in his diary, Marco Polo noted that burning of mineral oil gave light and heat, and Christopher Columbus used bitumen to make his ship seaworthy [1, 2].

But it was not until A.D. 1859 that exploration for oil and gas started in earnest, when the first oil well was drilled by Edwin Drake in northwestern Pennsylvania, United States (some quarters claim it started in 1846 in Azerbaijan). Since then, a lot of advancement took place in the field of oil exploration and production (E&P), and there has been a phenomenal growth in petroleum industry, making it one of the most important sectors in the world influencing global economy and life of the people across the planet.

1.2 E&P ACTIVITIES AND PROCESSES

Hydrocarbon E&P is a complex process beginning with prognostication and involving a series of activities, namely, geological survey, magnetic survey, gravitational survey, seismic survey, laboratory studies, geochemical study, and exploratory drilling encompassing coring, casing, cementing, mud engineering, and drill stem test (DST) followed by well testing. Based on the well testing results, the well is declared as "dry" or "hydrocarbon bearing." If no oil and gas are found, the well is abandoned. In case of discovery, another set of activities follow, namely, drilling of appraisal well, delineating of field, and assessing commercial viability of reserves. Based on these, the decision of the development of the field is taken; however, the scale of development is dependent on the potential of the field. Accordingly, field development plans are made and development wells are drilled; production installations and surface facilities (group gathering station (GGS), gas collection station (GCS), central tank farm (CTF), effluent treatment plant (ETP), etc.) are created before commencing production. All these activities are highly capital-intensive, and the gestation period for the realization of investment is quite long.

1.2.1 High-Risk and High-Cost Activity

The upstream oil and gas industry is unique. In conventional industry, inputs and outputs are deterministic, that is, with a given input (investment), one is assured of the planned output (product or services). But in the upstream oil and gas industry, the input is deterministic, but the output (outcome of exploration activity) is stochastic. With the planned investment, one is not sure about its realization—it's more like a gamble associated with uncertainty and high risk. More often than not, investment in exploration may not yield fruitful result or any return. Even if oil and gas are discovered, its commercial viability is to be assessed before the next course of action is decided. It takes a long time to develop the field before production begins. All these make E&P activities high-risk and high-cost operations.

1.2.2 High Technology Activity

Oil and gas E&P activities are technology-intensive and require expertise of diverse fields. E&P activities are essentially the application of various streams of science and engineering, such as science (geology, geophysics, geochemistry, palynology, mathematics, and statistics); engineering (petroleum, chemical, reservoir, mechanical, electrical, civil, marine and ocean, electronics, instrumentation, telecommunication, and computer science); and many others.

With depletion of easy reserves, E&P activities are becoming highly technology-intensive, as the search for oil and gas is directed to geographically and geologically difficult locations, such as deepwater exploration, arctic region, snowbound hostile terrains, mountains, deep oceans, high-pressure and high-temperature horizon, and other challenging areas. Moreover, with a phenomenal rise in global demand for oil and gas, future oil/gas production will mostly come from more difficult reservoirs, such as deeper horizon, low API gravity, and high sulfur content. Furthermore, the production of oil from aging field using conventional technology is a challenging task. The conventional technology too needs continuous improvement to sustain oil production from matured fields. *All these necessitate continuous development and induction of state-of-the-art technology, which are costly and require experts to use it and make the best out of it.* E&P activities are associated with high technology that requires multidisciplinary approach and expertise to operate "state-of-the-art technology" and cope with increasing demand and difficulties in oil and gas E&P.

1.3   NEED FOR OPTIMIZATION IN UPSTREAM INDUSTRY

We have seen in the earlier paragraph that oil and gas E&P activities are becoming increasingly costly, risky, and technology-intensive as operations are moving from easy to difficult and challenging frontiers. In order to mitigate risks and share the cost of operations, even the major and super major oil companies are forming joint ventures and consortium for venturing in new frontiers. In view of inherent risks and uncertainty associated with the upstream business where inputs are deterministic but output is probabilistic, it is important that oil companies use their capital and resources judiciously. It is necessary to optimize strategies, resources, and cost and improve business performance in all spheres of E&P activity. These are the need for survival and sustaining business. The rule of the game is "money saved is money earned." All these require innovative ideas, change in mind-set, fresh outlook, and approaches to business.

1.3.1   Optimization Techniques

Optimization is an oft-repeated word used by all, whose meaning perhaps is not as clear as it seems to most of the people. It's a catchy word! People use it liberally, as it sounds impressive without knowing its nuances or relevance to the context. Most people consider it as a synonym of maximization/minimization, and the differences are indistinct even to professionals and management people.

Optimization in its simplest form means the best available value or most favorable result *under a given set of conditions or constraints. It is usually the* maximization or minimization *of objective function subject to a set of constraints.* Optimization is basically a mathematical technique, which is widely used in engineering, management science, economics, science, mathematics, and many other fields. Literature is replete with definition of optimization with varying degree of simplicity or complexity.

The genesis of "optimization technique" traces back to the work of Fermat and Lagrange for identifying optima with calculus-based formula. Newton and Gauss used iterative methods for moving toward an optimum solution. In modern days, George B. Dantzig developed an optimization technique called "linear programming" based on simplex algorithm. It was developed during World War II for scheduling warfare logistics and related problems for US military. Much of the work of G. B. Dantzig was based on the theory introduced by Leonid Kantorovich in 1939, but Dantzig made substantial improvement on it making it more powerful and versatile [3]. Based on the types of objective function and set of constraints, the optimization models/techniques are classified as linear programming, integer programming, geometric programming, goal programming, quadratic programming, nonlinear programming, fractional programming, dynamic programming, and so on. These are essentially the extension of either linear programming or particular case(s) of nonlinear programming.

Various optimization and business improvement techniques have been used in this book, such as benchmarking, technical and qualitative analysis to optimize productivity of drilling operation (Chapter 2); diagnostic approach and root cause analysis to optimize controllable rig time loss (Chapter 3); technical, qualitative, and economic analysis to optimize geology and geophysics (G&G) strategy for deepwater oil and gas exploration (Chapter 4); queuing theory to determine optimum number of offshore supply vessel (OSV) fleet size (Chapter 5); technical and statistical analysis for standardizing consumption of consumables in oil/gas wells and rigs (Chapter 6); critical path analysis using Program Evaluation and Review Technique/Critical Path Method (PERT/CPM) to optimize rig move/mobilization time and activity scheduling (Chapter 7); development of uniform standards for emergency alarm systems and indicators at offshore installations based on recognized international codes (Chapter 8); qualitative and quantitative analysis to optimize supply chain management (SCM) system (Chapter 9); best practice benchmark, work study, qualitative and quantitative analysis for manpower optimization, and strategic workforce planning (Chapter 10); enhancement of organizational efficiency through business process simplification (Chapter 11); and linear programming to optimize base oil price (Chapter 12).

1.4   IMPORTANCE OF CREATIVITY AND DATA USABILITY FOR BUSINESS PERFORMANCE IMPROVEMENT

E&P companies usually maintain a plethora of operational data and use these in good measure for preparation of reports, monitoring of activities, review, and decision making. Apart from these, the huge caches of data mostly remain in archive with limited usability, but this can turn to a treasure trove, if dealt with innovatively. What's needed are creative ideas, fresh outlook, and analytical abilities, which are often stifled in this fast-paced business world. Although there is no dearth of talent but work pressure, tight schedule for delivery, annual commitment, and so on, don't leave much room for creativity to flourish. A good idea or groundbreaking study need not necessarily be complex; in fact, most of them spring from simple ideas. A good idea or powerful study is easy to understand and easy to implement and brings in desired improvement. The characteristics of a good idea or innovative study are that it looks simple and is easily understood by most of the people, yet no one conceived or figured it out until it was presented. Like no one bothered till Sir Isaac Newton explained—"why apple falls on the ground." That's the hallmark of creativity!

1.5   OVERVIEW OF THE BOOK

As the title suggests, this book deals with optimization and business improvement studies in the backdrop of upstream oil and gas industry, but some of these studies, approaches, and techniques are also applicable to other industries. The book contains studies on optimization of strategies, resources, and cost; improvement in business performance and operational productivity; identification and removal of inefficiencies; standardization of consumption of materials; standardization of important safety measures; business process simplification, manpower optimization, improvement in human resource productivity, and so on. Various business processes, systems, and operational areas in E&P business were studied, inefficiencies were identified, and measures for improvement were suggested. The purpose of the book is not to delve deep into the operational technicalities but to emphasize on the approach for optimization and improve operational and functional performance using quantitative and qualitative tools. Therefore, technical discussions related to operational and functional areas have been kept at necessary level.

The book is divided into 12 chapters; besides the introductory chapter, it contains 11 real-life optimization and business improvement studies that are worth mentioning and emulating. Chapter 1 is the introductory part that explains the purpose and structure of the book. It portrays the overview and chapterwise contents of this volume covering a wide spectrum of activities from E&P operations to business process improvement and the like.

In order to contain the spiraling cost of drilling, especially at offshore, the E&P companies and drilling operators are continuously trying to improve the productivity of drilling operation, which is a necessity for survival in these days. Chapter 2 deals with optimization of productivity of drilling and dispels few long-held beliefs in the organization about poor performance of own rigs compared to the hired rigs. The study diagnoses areas of concern, identifies major factors affecting drilling performance, and categorizes these as *human factors*, *organizational factors*, and *technical factors*. All these factors and subfactors were analyzed, and remedial measures were suggested for optimizing the productivity of drilling operations with the potential of saving around USD 60.5 million per year for offshore Asset under study.

Drilling is a capital-intensive activity consuming a lion's share of the capital budget of an E&P company. Therefore, it is desirable to minimize nonproductive drilling time and improve rig time availability. Chapter 3 discusses optimization of controllable rig time loss using diagnostic approach. The study identifies causes of rig time loss, quantifies, and groups these under five categories, namely, waiting on material, waiting on decision, waiting on logging tool, equipment repair downtime, and other shutdown. The study reveals that controllable rig time loss accounts for 14.2% of available rig-days costing around USD 60.5 million per year in the Asset under study. Remedial measures to minimize these losses have been suggested, which would entail a saving of around USD 34.5 million per year in the Asset under study.

Deepwater exploration holds promising prospects for the future, as easy oil and gas reserves are depleting fast in this oil-hungry world. But deepwater exploration is costly, risky, and technologically challenging, which necessitates extreme economic prudence and strict monitoring of E&P operations. Ironically, some G&G activities escape attention because of inherent subjectivity and uncertainties involved in these operations. Chapter 4 aims to optimize G&G strategies for exploring oil and gas in deepwater and analyze some G&G decisions and their effect on well economics through the following: (i) optimization of G&G evaluation time in deepwater wells, (ii) techno-economic assessment of acquiring logging while drilling (LWD) in own deepwater rig, (iii) improvement of accuracy of geological predictions, (iv) effect on downhole complications due to variation in formation pressure (between actual and predicted pressure in a well), (v) containment of slippage in deepwater well completion, and (vi) influence of people's factors on the success and performance of deepwater exploration. The study provides valuable insights and offers suggestion in these areas, which would help in optimizing G&G strategy in deepwater exploration with the potential of saving around USD 50 million in less than a year.

OSV is the lifeline of offshore E&P operations. Waiting cost of offshore installations is extremely high, and it is undesirable that installations wait for materials that are supplied by OSVs. Therefore, adequate number of OSVs is required for uninterrupted operation, but OSVs are also costly items. It is necessary to trade-off between waiting time of installations and that of OSVs. Chapter 5 develops a queuing model to optimize OSV fleet size with the objective of minimizing waiting time of installations, which in turn would lead to optimizing waiting cost of installations and the total system cost. The study also determines various operating characteristics of the queuing model, which would help in decision making under dynamic conditions.

Chapter 6 deals with standardizing consumption of high-speed diesel (HSD), cement, and chemicals in oil/gas wells and rigs to prevent stockout situation, avoid excess inventory, monitor consumption, check wastages and aberrations, and help drawing future procurement plan. HSD consumption depends on various technical, geological, and physical factors. **Average HSD consumption** per meter drilling for similar category and capacity of rigs operating in similar geological formation and depth with similar rate of penetration was grouped for standardization. Accordingly, unit fuel consumption for different types and capacities of rigs in different regions/Assets/Basins were computed and standardized both at onshore and offshore. Suggestions have been offered for the improvement of HSD consumptions in drilling rigs, which can save up to USD 62.8 million per year. **Consumption of cement** depends on various technical and geological factors like casing policy, hole size and depth, cement rise, number of objects to be tested, activity and mud loss, downhole complications, and so on. Consumption of cement was standardized for different casing policy wells in different fields/regions taking into consideration the aforementioned factors. **Consumption of chemicals** in an oil/gas well depends on various technical and geological factors, such as well depth, formation pressure, lithology, and borehole instability, to mention a few. The unit consumption (kg/m) of chemicals in oil/gas wells varies widely with high standard deviation, not only in different fields but even within the same field. Therefore, consumption range and upper limit of consumption of chemicals have been determined for various groups of wells.

Rig move/mobilization is considered as unproductive rig time, which needs to be minimized to ensure more time for drilling and completion. Network analysis using PERT/CPM technique has been used in Chapter 7 to optimize rig move time and develop optimal activity schedule. The study identifies critical path and critical activities and focuses on timely completion of critical tasks to avoid delay in rig move/mobilization. The study suggests a set of recommendations for optimizing rig move time and activity schedule, which has the potential of saving 500 rig-days amounting to USD 30 million per year in the E&P company under study.

Chapter 8 focuses on developing uniform standards for emergency alarm systems and code of signals for offshore installations of an E&P company, which are found to vary widely. This creates confusion and possesses safety threat to offshore-bound personnel, especially those who frequent different installations. The study classifies different emergency situations, and provides useful guidance for developing code of signals for emergency alarm system and indicators based on recognized international codes. It suggests uniform adoption of standards across various offshore installations of the organization.

Chapter 9 deals with optimization of SCM system and highlights importance of SCM system for smooth functioning of E&P operations. It identifies opportunities for improvement by assessing maturity of key supply chain functions and benchmarking these with comparable industry standard. The study aims to improve procurement cycle time by streamlining material planning, tendering, and order execution processes along with other measures. It also suggests improvement in inventory management and warehouse functions. Furthermore, the study emphasizes the need for SCM support services, namely, demand forecasting, strategic procurement, and vendor relationship management, which are currently unorganized or nonexistent in the organization under study. The recommendations made in this study can improve SCM system in the organization and have the potential of saving USD 244 million per annum.

Chapter 10 discusses manpower optimization and strategic workforce planning of an E&P company emphasizing on multiskilling, multidisciplinary approach to improve employee utilization and productivity and enrich the quality of human resources. It rationalizes the large pool of disciplines/subdisciplines and impresses on modification of the organization's manning norms aligning with the best practice benchmark. It studies the current manpower planning process in the organization, finds out shortcomings that are not conducive to good practices, and suggests measures to overcome these. A real-life case study has been presented illustrating the process of manpower optimization that includes demand forecasting, supply (availability) prediction, and balancing demand–supply gap. The study reveals that there is enormous scope for savings on account of manpower optimization in the organization, approximately USD 290 million.

Chapter 11 illustrates some real-life examples of business process simplification, which is a powerful tool to improve system efficiency, especially in large enterprises where processes are deeply embedded in organizational structure and culture and seem inseparable. The study highlights the efficacy of business process simplification in improving customer satisfaction and service quality, reduction of processing time, elimination of nonvalue adding tasks, and duplication of efforts, freeing up scarce resources and creating awareness in an enterprise.

Oil pricing is a complex and sensitive issue, which is not dependent on economic criteria alone—a set of environmental factors like social, political, technical, and geopolitical issues strongly influence it. Chapter 12 develops a quantitative model to optimize base oil price for a country using linear programming, taking into account the cost and share of domestic oil production and that of oil import and other factors. It aims to maximize profitability and drive oil production by reinvesting into E&P activities. It is an illustrative model that develops a framework to study the effect of various parameters on base oil price and aid decision making under dynamic circumstances.

All these are real-life examples of optimization and business improvement studies, which are based on simple concept, notwithstanding inherent uncertainties involved in E&P business. A wide range of topics are covered in the book containing powerful drivers to capitalize opportunity cost and shore up business performance. These studies are easy to emulate and have the potential of saving billions of dollars, besides improving organizational efficiency.

REVIEW EXERCISES

1.1 Describe the importance of oil and gas in today's world.

1.2 What are the sequence and process of E&P activities?

1.3 Is there any need for optimization in upstream industry? Explain.

1.4 Define optimization. Is it different from maximization/minimization? Explain.

1.5 What are the various optimization techniques you are aware of? Mention their applicability.

REFERENCES

[1] Chowdhury, S., Striking Oil, Science Reporter, Volume 31, Number 5, May 1994, pp. 9–12, 35.
[2] Mee, A., Children's Encyclopedia, Vols. 3 and 4, Everybody's Publications Ltd., The Educational Book Company, London, 1963.

Useful Link

[3] Wikipedia, Mathematical Optimization, available at: http://en.wikipedia.org/wiki/Mathematical_optimization (accessed on February 27, 2016).

FURTHER READING

Campbell, J.M., Campbell Petroleum Series, Analysis and Management of Petroleum Investments: Risk, Taxes and Time, 2nd edition, PennWell Publishers, Tulsa, OK, 1991.

Campbell, J.M., Campbell, C.W., Campbell, R.A., Friedman, G. Campbell Petroleum Series, Mineral Property Economics, Vols. 2 and 3, The Campbells, Tulsa, OK, 1980.

Hobson, G.D. and Tiratsoo, E.N., Introduction to Petroleum Geology, Gulf Publishing Company, Houston, TX, 1981.

Mohan, C. and Deep, K., Optimization Techniques, New Age Science, Kent, 2009.

North, F.K., Petroleum Geology, Allen and Muir, Inc., Winchester, MA, 1985.

2

OPTIMIZING PRODUCTIVITY OF DRILLING OPERATIONS

2.1 INTRODUCTION

Drilling is an expensive and technology-intensive activity, which consumes a major part of an E&P company's budget, especially in offshore Assets and Basins. The cost of drilling is increasing over the years throughout the world, which is confirmed by various reports and studies conducted by well-known groups, agencies, authorities, and organizations like Joint Association Survey [1], US Energy Information Administration (EIA) [2], E&P companies, and so on. This is expected due to the fact that easy and shallow reserves are depleting fast, and the search for and production of oil and gas are directed to greater depth and in difficult areas (e.g., production from matured fields using secondary and tertiary recovery methods, exploring in ultradeep water, producing from deep reservoir, targeting small and marginal oil pools from the same well). This requires acquisition and use of advanced technology, which adds to the cost. In addition, the cost of drilling increases exponentially with depth, which is also confirmed by various E&P companies and studies carried out by different groups and experts [3, 4].

Furthermore, E&P activities have increased over the years with the rise in demand for oil. All these have resulted in higher demand for rigs. But the manufacturing and supply of rigs have not kept pace with the soaring demand. As a result, there is shortage of drilling rigs all over the world, thus pushing up rig hiring cost. All these contributed to an increase in drilling cost over the years. E&P companies and drilling contractors are striving hard to contain the rising drilling cost through various means under their control like efficient use of available resources and economic prudence. They are always on the lookout to find ways to minimize drilling cost and improve drilling productivity.

The terms "drilling efficiency" and "drilling productivity" are used interchangeably in E&P industry, although there are subtle differences. Efficiency is a metrics of output for

Optimization and Business Improvement Studies in Upstream Oil and Gas Industry, First Edition.
Sanjib Chowdhury.

a given set of inputs; it highlights how best the resources are used for a specific output. Productivity, in simple terms, is the rate of output with respect to inputs conforming to standards or quality. Drilling efficiency is generally measured in terms of footage drilled per day or per rig, resources added per well or per rig, wells drilled per rig, drilling cost per foot, energy consumption, drilling speed, and so on. The term "drilling productivity" has wide connotations and is interwoven with various intricate issues, such as operational productivity, geologic conditions and reservoir properties, drilling parameters, achieving target objectives, economic and financial factors, and so on. Different groups usually focus on different aspects mentioned previously depending on their expertise, interest, capability, and objectives. A concerted effort is required to assimilate related information and knowledge for cost optimization and improving the productivity of drilling.

It may be noted that a common benchmark for "drilling productivity" would be unrealistic due to wide variability and large number of factors affecting drilling performance. Therefore, it would be appropriate to benchmark drilling performance under similar geologic conditions and depths; otherwise, it would be an inappropriate comparison. *This study deals with the improvement of "drilling productivity" taking into account many of the aforementioned factors. It benchmarks drilling performance of own drilling rigs against the contract rigs operating in the same area under similar geological conditions. The factors affecting low performance were identified and analyzed, and measures for improvement were suggested.*

Before we proceed further, a brief note on drilling operation and cost is explained in the next paragraph, which may be useful to the esteemed readers.

2.2   A BRIEF NOTE ON DRILLING OPERATION AND COST

Drilling technology has evolved over the years—drilling in earlier days was percussive type, that is, raising and dropping cable tool with a pulley system. But, major changes took place with the introduction of rotary drilling in 1930s, which paved the way to drill more efficiently to a greater depth and at a faster rate. Drilling rig is a common landmark of upstream oil and gas industry, which is easily recognized by the huge and tall steel structure called "derrick" anchored to the ground. Drilling rig consists of a number of components and equipment, such as derrick, high-powered engines, mud pump, mud circulation system, draw works, blowout preventer, water tanks, fuel tanks, and so on (Fig. 2.1). The traveling block is suspended from the crown block—a large pulley at the top of the derrick. The swivel attached by a large hook to the traveling block can rotate freely, and the Kelly is fitted onto this. A rotatory table at the center of the derrick floor holds the Kelly (which has a square or hexagonal cross-section) and can be rotated at a desired speed by the engine. To begin drilling operation, the Kelly is hauled up the derrick, and its bottom is fitted with a drill bit and lowered through the rotary table until the bit rests on the earth. With the start of engine, the rotary table rotates the Kelly and drill bit, which is pressed hard against the earth by the weight of the drill string, thus cutting and penetrating the rock at the bottom [5].

Drilling fluid, popularly known as mud, is a specially designed slurry of chemicals, salt, and water, which is run through the drill pipe to the bottom of the hole. It comes out from the nozzle of the drill bit and makes way to the surface through annulus of the drill pipe and borehole wall/casing. It serves several important functions, such as flushing out of rock cuttings, cooling and lubricating the drill bit, balancing formation pressure, and so on.

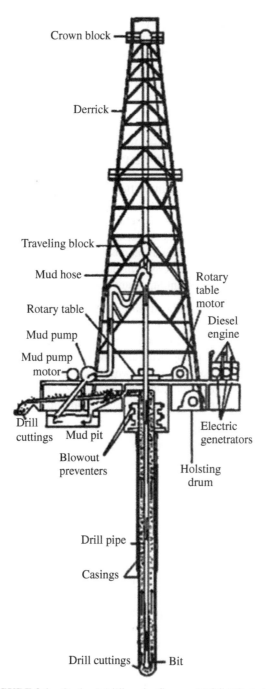

FIGURE 2.1 On-land drilling rig. Source: NISCAIR, India.

As the well gets deeper, additional drill pipe is connected with the drill string under the Kelly or top drive at the surface, which is known as "making a connection" or tripping. After drilling a certain depth, the casing pipes are lowered and cemented to prevent collapse of the sidewall and retain structural integrity of the well, which are known as casing and cementing. The wellbore is telescopic in nature whose ultimate productive size depends on the number of casing string. While drilling a well, several operations, namely, well logging, coring, production testing, and so on, are carried out to evaluate the formation properties. All these important functions, namely, mud circulation and drilling fluids, casing and cementing, and formation evaluation—logging, coring, production testing, and so on, are further elaborated in Chapters 4 and 6.

Till 1970s, oil and gas wells were mostly vertical; the development of directional drilling took place in 1970s, which culminated in the development of horizontal drilling. Nowadays, horizontal drilling is a common practice, which has many advantages over vertical wells and is especially useful: (i) for drilling multiple wells from single location (e.g., central offshore platform) targeting different reservoir zones, (ii) horizontal wells have greater surface area with production zone (i.e., maximum reservoir contact) in thin reservoir than vertical wells resulting in higher flow rates which in turn make it economic, and (iii) inaccessible location, for example, densely populated area, environmentally sensitive area, drilling offshore location from onshore, and so on. The advent of top drive system (TDS) replacing the Kelly drive system vastly improved horizontal well capability due to its inherent advantages of reaming up and down and reducing connection time.

Since then, there has been remarkable advancement in drilling technology; sophisticated tools and equipment like measurement-while-drilling (MWD), logging-while-drilling (LWD), and so on, provide real-time information about subsurface properties and drilling parameters. All these help drilling engineers and G&G professionals to take quick decision and follow safe course of action to avoid downhole complications and improve drilling efficiency. Multilateral drilling technology that enables drilling multiple horizontal wells from the same well vastly improved cost and resource optimization.

Based on areas of operation, drilling activity is called on-land drilling and offshore drilling. The drilling process and equipment are mostly same for both on-land and offshore drilling, except the types of rigs used. The difference between on-land and offshore rig is mainly the way rig is supported. The design and engineering of rig support for offshore is far more complex and different from that of on-land rigs. Due to marine environment and remoteness from the shore and base, offshore rigs require additional support, such as diver support, logistics (marine and aviation) support, meteorology station, and so on.

Depending on water depth, offshore rigs are either floating type (e.g., drill ship and semi-submersible), moored or unmoored type (dynamic position rig), or fixed to the seabed (e.g., jack-up). Offshore drilling rigs are mainly of three types—Jack-up (suitable for drilling up to 500 ft water depth); platform type, semisubmersible (capable of drilling at 1,650–9,900 ft water depth); drill ships (capable of drilling at 1,650–12,000 ft water depth). Semisubmersible rig and drill ships may be anchored or dynamic-positioned. The capability of anchored rig is limited to water depth 5,000 ft, whereas dynamic-positioned drill ship is capable of drilling up to water depth 12,000 ft.

The hiring charges of drilling rig is a major component of well cost. It accounts for nearly **30% of well cost and half of rig operating cost at offshore** (refer to Section 2.4g, Figs. 2.10 and 2.11). The current (2015) rig hiring charges of offshore drill ship capable of drilling above 4000 ft water depth is approximately USD 500,000 per day, and those less than 4000 ft water depth is approximately USD 250,000 per day. Similarly, hiring cost of semisubmersible rig capable of drilling above 4000 ft water depth is approximately

USD 440,000 per day, between 1500+ and 4000 ft water depth is approximately USD 350,000 per day, and below 1500 ft water depth is approximately USD 300,000 per day; jack-up rigs of varying capacity and type for water depth less than 200 ft to more than 300 ft range approximately between USD 90,000 and 180,000 per day [6].

Similarly, on-land rigs are of different types depending on capacity, maneuverability, and need, namely, (i) mobile rig, (ii) desert rig on wheel, (iii) rigs with high floor mast with substructure, and (iv) combination thereof. On-land rigs are also classified as shallow, medium, deep, and superdeep based on drilling capacity and depth. On-land rigs are comparatively cheaper than offshore rigs. The hiring cost of on-land rigs varies widely depending on the type and capacity of rigs. Typical on-land rig hiring cost ranges between USD 25,000 and 40,000 per day at present.

Drilling efficiency is dependent on a large number of factors, such as geological conditions and lithology of the field (i.e., rock composition, hardness, and dip of beds), drilling parameters (bit design and type, rate of penetration, mud circulation, etc.), availability of tools, equipment and technology, rig efficiency, operating experience, and others. To learn more about oil well drilling, interested readers may refer to "References/ Further Reading" listed at the end of this chapter.

2.3 OBJECTIVES

This exercise was carried out in a large offshore Asset of an E&P company having its own fleet of drilling and workover rigs with full-fledged crews and in-house services, such as drilling and cementing services, well services, logging services, logistics services, and so on. Hired services are also availed depending on need and work program. Drilling operation is also performed by contract rigs, as in-house availability of drilling rigs and resources is not sufficient to meet the work program and future need.

It has been generally observed that the performance of own rigs and crews is not at par with that of contract rigs. There is considerable scope for improvement in this regard. In order to bridge this gap, the objectives of this study are as follows:

(a) To identify key factors for low performance of own drilling rigs and crews
(b) To benchmark key performance indicators of drilling and related operations
(c) To suggest remedial measures and *optimize productivity of drilling operation.*

2.3.1 Methodology

The methodology followed in this study is described in the following paragraph:

(a) Performance of drilling and related activities in a large offshore Asset was studied for a period of 2–3 years. The data of approximately 120 wells drilled during this period were critically examined.
(b) Performance of various activities and related parameters like cycle speed, commercial speed, well completion, and time slippage of both own rigs and contract rigs were compared over a period of 2–3 years.
(c) Performance data was incisively analyzed and discussed with key officials, and the reasons for low performance were identified.

(d) Finally, remedial measures were suggested to optimize the productivity of drilling operation.

2.4 KEY OBSERVATIONS AND FINDINGS

The key performance indicators of drilling operation, such as cycle speed, commercial speed, well completion, drilling time slippage, production testing, and so on, and also rig vintage/profile, well cost, and so on, were analyzed. The findings are presented in the following paragraphs:

Performance Indicators

(a) Cycle speed is a drilling performance parameter and is defined as the ratio of length or depth drilled to cycle time. It is expressed as feet or meterage drilled per rig-month. Cycle time is the time from rig release at one well to rig release at the next well.

Cycle speed of hired rigs is 57%, 85%, and 58% higher than that of own rigs during Year 1, Year 2, and Year 3, respectively (Fig. 2.2).

(Year 1: own—765, hired—1197; Year 2: own—689, hired—1278; Year 3: own—803, hired—1268 m/rig-month).

(b) Commercial speed is a measure of drilling performance and is defined as the ratio of length or depth drilled to actual drilling time. It is expressed as feet or meterage drilled per rig-month. Commercial time is the time between rig spud date and well completion date.

The commercial speed of hired rigs is found to be 36%, 77%, and 76% higher than that of own rigs during Year 1, Year 2, and Year 3, respectively (Fig. 2.3).

(Year 1: own—1097, hired—1486; Year 2: own—857, hired—1516; Year 3: own—863, hired—1520 m/rig-month).

(c) The number of wells completed per rig-year by hired rigs are 53%, 90%, and 50% higher than that of own rigs during Year 1, Year 2, and Year 3, respectively.

(Year 1: own—3.6, hired—5.5; Year 2: own—2.8, hired—5.3; Year 3: own—3.7, hired—15.6; Fig. 2.4).

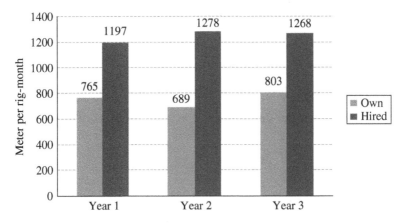

FIGURE 2.2 Comparison of cycle speed.

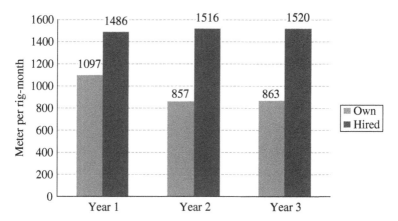

FIGURE 2.3 Comparison of commercial speed.

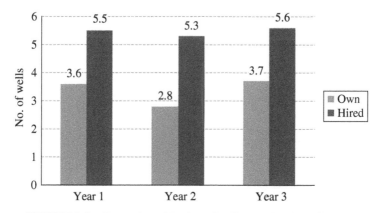

FIGURE 2.4 Comparison: Number of wells completed per rig-year.

(d) Well completion time slippage is defined as the ratio of excess time taken to complete a well to planned well completion time. That is, (actual well completion time – planned well completion time)/planned well completion time.

   Time slippage (%) over planned well completion time is considerably higher in own rigs than hired rigs during Year 1 and Year 2 (Fig. 2.5).

   (Year 1: own—59.3%, hired—28.2%; Year 2: own—63.6%, hired—7.4%).

(e) Similarly, production testing time overrun is the ratio of excess time taken for production testing to planned production testing time. That is, (actual production testing time – planned production testing time)/planned production testing time.

   Time overrun for production testing is considerably higher in own rigs than contract rigs during Year 1 and Year 2 (Fig. 2.6).

   (Year 1: own—56.7%, hired—15.6%; Year 2: own—106.7%, hired—6.8%).

Rig Profile (Vintage)
There was a common perception in the company that own rigs and equipment are much older than that of contract rigs resulting in low performance of own rigs compared to

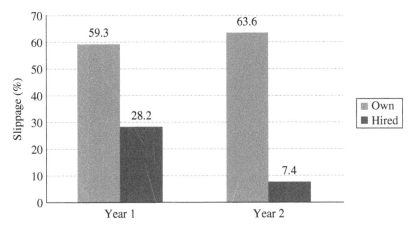

FIGURE 2.5   Well completion time slippage (%).

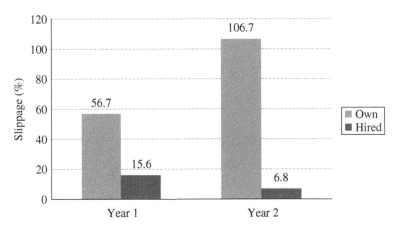

FIGURE 2.6   Comparison of production testing time slippage (%).

contract rigs. On examining, this perception was found untrue. The study revealed many startling facts, which are presented in the following paragraphs:

(f) The average age of contract rigs is 6.2 years higher than own rigs (Fig. 2.7). But the performance of hired rigs in terms of cycle speed, commercial speed, number of wells completed per rig-year, and so on, is consistently and considerably higher than own rigs.

The average age of own rigs is 19.3 years, and their age varies between 13 and 22 years, except one (rig: own 10), which is 30 years old (Fig. 2.8).

On the contrary, the average age of hired rigs is 25.5 years, and their age varies between 20 and 30 years (Fig. 2.9).

These findings are contrary to the long-held belief of the employees that own rigs are much older than contract rigs.

Well Cost Analysis
The wells drilled in the last 2 years were analyzed and the outcome is presented in the following paragraphs:

FIGURE 2.7 Average age of own and hired rigs (as on 12/20xx).

FIGURE 2.8 Age of own rigs (as on 12/20xx).

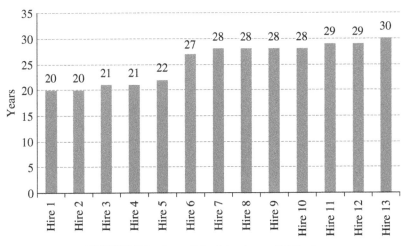

FIGURE 2.9 Age of hired rigs (as on 12/20xx).

(g) The major cost components of wells are found as rig operating cost—57%, logging—13%, casing pipes—8%, cementing—6%, rig move—3%, and others—13% (Fig. 2.10).

The components of rig operating cost are found as rig hiring and mobilization—48% (i.e., ~30% of well cost), offshore supply vessels (OSVs)—20% (i.e., ~12% of well cost), stores and spares—7%, POL—6.2% (i.e., ~3.8% of well cost), manpower—5%, and others—14% (Fig. 2.11).

(h) Predictably, the deployment of hired rigs is costly than that of own rigs. The average rig operating cost per day of hired rigs is 80% and 92% higher than that of own rigs in Year 1 and Year 2, respectively (Fig. 2.12).

(Year 1: own—USD 39,500, hired—USD 71,250; Year 2: own—USD 35,000, hired—USD 67,250).

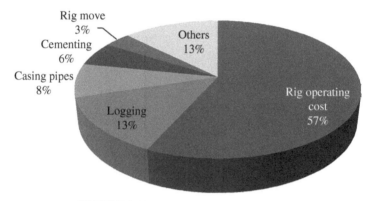

FIGURE 2.10   Major well cost components.

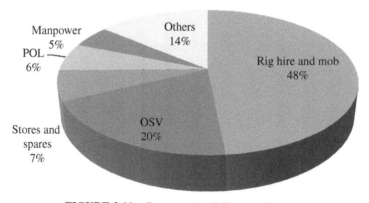

FIGURE 2.11   Components of rig operating cost.

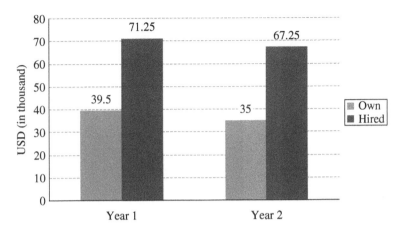

FIGURE 2.12 Comparison of average rig operating cost per day.
Note: The average cost on POL, OSV, helicopter, drilling overheads, contractual, and IMR jobs is nearly the same at USD 17,500 per day for both own and hired rigs. The average rig hiring and mobilization charges are around USD 50,000 per day. The average cost for spares, catering, manpower, repair and maintenance, insurance, and depreciation at own rigs is around USD 17,500 per day.

2.4.1 Analysis of Key Observations

The study dispelled the long-held belief of the organization that aging drilling rigs and equipment were responsible for poor performance of own rigs compared to hired rigs. On the contrary, it was observed that the average age of contract rigs is 6.2 years higher than own rigs. The company officials knowingly or unknowingly used this as a defense for low performance of own rigs.

The analysis of key performance indicators, such as cycle speed, commercial speed, well completion time, slippage in drilling and well completion time, and so on, reveals that the performance of hired rigs is consistently better than own rigs despite the fact that hired rigs are older than own rigs. The main reasons for low performance of own rigs and crew can be broadly categorized and attributed to the following:

  (i) *Human factors*
  (ii) *Organizational factors*
  (iii) *Technical factors*

Further details on these factors are discussed in the following paragraphs.

2.5 HUMAN FACTORS

Human factors include rig crew composition and staffing (i.e., age, quality, and strength) and competency (knowledge, skills) of the crew. This section has two subsections, namely, (i) areas of concern and (ii) suggestions for improvement.

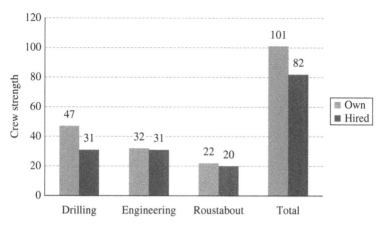

FIGURE 2.13   Comparison of crew strength at own and hired rigs.
Note: Engineering crew includes staff of mechanical and electrical disciplines, crane operator, welder, and storekeeper.

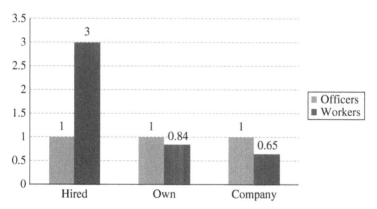

FIGURE 2.14   Officers: workers ratio at own and hired rigs.
Note: Assuming job positions in hired rigs are similar to own rigs, that is, rig superintendant, tool/ tour pusher, driller, assistant driller, chief mechanic, and chief electrician are equivalent to officers; derrick man, rig/floorman, motorman, electrician, crane operator, welder, roustabout, and storekeeper are considered as workers (highly skilled).

2.5.1   Areas of Concern

The areas of concern pertaining to the aforementioned human factors are discussed in the following paragraphs:

Rig Crew Composition and Staffing

   (a) The number of crew deployed at own rigs are 23% higher than that of contract rigs. The difference is mainly in the staffing of drilling crew (own—47, hired—31; Fig. 2.13).

      Despite the advancement of technology, the staffing pattern, work culture, and work practices remain unchanged in own rigs. These are true for both onshore and offshore Assets of the company.

   (b) The ratio of officers to workers posted at own rigs is 1 : 0.84 compared to 1 : 3 at hired rigs (Fig. 2.14).

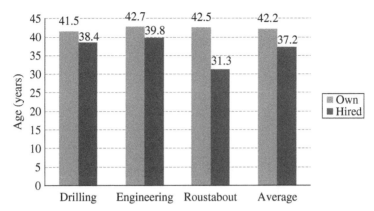

FIGURE 2.15 Comparison of crew age at own and hired rigs.

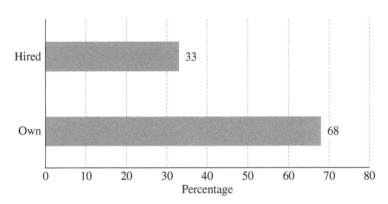

FIGURE 2.16 Crew above 40 years of age (%).

There are more officers than workers even in fields and operations! The situation is no better in office and company, as a whole. Such adverse officers' to workers' ratio is rarely seen in other E&P companies and is not a healthy sign. It points toward a slack career progression policy that is not aligned with the best practices and requires attention.

Age Profile

(c) The average age of crew working at own rigs is 5 years older than that of contract rigs (own rigs—42.2 years, hired rigs—37.2 years; Fig. 2.15).

The average age of roustabout posted at own rigs are 11.2 years higher than that of hired rigs (own rigs—42.5 years, hired rigs—31.3 years).

The youngest roustabout in own rigs is 33 years compared to 18 years in hired rigs. Even the age of the youngest roustabout (i.e., 33 years) in own rigs is higher than the average age of roustabout (31.3 years) in contract rigs. Similarly, the oldest roustabout at own rigs is 59 years compared to 56 years in hired rigs.

(d) A total of 68% of crew working at own rigs are above 40 years of age compared to 33% in hired rigs (Fig. 2.16).

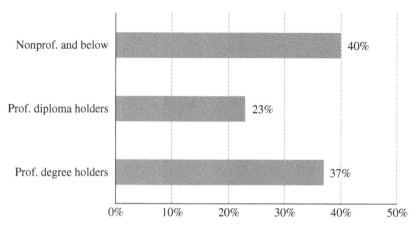

FIGURE 2.17   Qualification profile of officers posted at rigs.

A total of 26.5% roustabouts in own rigs are above 45 years of age compared to 4% in contract rigs. Similarly, 10.3% roustabouts at own rigs are above 50 years of age as against 1.6% at contract rigs.

The aged crews are not suitable for such laborious work, thus affecting performance of drilling operation. These are serious issues as the job of roustabouts is highly laborious and considered as hard duty.

Competency: Knowledge/Skill Base

(e) The knowledge/skill base of officers posted at drilling rigs is as follows:

Professional (technical/engineering) degree holders—37%

Professional (technical) diploma holders—23%

Nonprofessional degree/diploma holders and below—40% (Fig. 2.17)

Around 40% of officers posted at offshore rigs do not have relevant qualification, knowledge, or skill base and have risen from the rank. Many of them are from nonrelated trades/fields like painting, turner, auto, draftsmanship, and so on.

Many of these technical officers are nonmatriculate (below 10th Class), Secondary School Certificate (SSC) passed, Trade Training Certificate (TTC) certified fitter, motor mechanic, diesel mechanic, auto mechanic, welder, and so on.

2.5.2 Suggestions for Improvement

Based on the findings and discussions in Section 2.5.1, the following suggestions pertaining to rig crew composition, staffing, age, and competency would help in optimizing the productivity of drilling operation.

Rig Crew Composition and Staffing

(a) The staffing pattern of rigs may be looked afresh keeping in mind the following:
   • Advancement of technology

- Scope of outsourcing of services
- Standard norms or practices followed by international drilling companies

To start with, the staffing pattern of hired rigs may be considered for adoption in own rigs.

Age Profile

(b) Considering the critical nature of some operations and intense manual labor involved in the deep drilling rigs, the average age of crane operator, welder, rig-man, and roustabout should be ideally around 35 years, with maximum up to 45 years. These positions can be considered for outsourcing or taken on contract for better work efficiency.

Competency: Knowledge/Skill Base

(c) Employees with skills of nonrelated trades may not be posted at the deep drilling rigs. Ensure that people with appropriate trade skills are posted at drilling rigs, especially in capital-intensive offshore rigs.

(d) Knowledge and skill competence gap of each trade/group, especially those involved in operations and maintenance work, may be assessed and mapped.

Effective measures may be taken to bridge these gaps through planned induction, development of existing people through special training, and posting of right people.

2.6 ORGANIZATIONAL FACTORS

Organizational factors include HR planning and policy covering induction, training and development, career progression, outsourcing, motivational aspects, discipline-related issues, food habits, and health of crew. This section is divided in two subsections, namely, (i) areas of concern and (ii) suggestions for improvement.

2.6.1 Areas of Concern

The areas of concern identified for the organizational factors are discussed in the following paragraphs:

HR Planning and Policy: Induction, Training and Development, Outsourcing, and Promotion

(a) There is lack of planning for replacement of aging population, induction at the entry level, proper training, and development of in-house personnel, especially in the critical categories, trade and disciplines.

(b) The career progression policy of the organization is rather loose and has been compromised due to various reasons. For example, a worker becomes an officer after certain years of service even though he/she does not have the basic qualification for the post. There has been a gradual erosion of HR policy and system in the organization over the years. One of the main reasons for this is inept HR administration, which often succumbed to the pressure of the trade unions (which are usually patronized by powerful political parties).

(c) There is hardly any inter-rig movement of crew. The workmen/crews continue to be attached with the same drilling rig for decades, even after they are promoted to the officer level.

Motivational Aspects

(d) Usually, symptoms are good indicators for assessing morale of employees in an organization. The following symptoms prevailing across the organization indicate not so high employee motivation:

- Maintenance and auxiliary staff prefer posting at production installations and platform rather than drilling rigs due to hard working conditions on rigs. Financial incentives do not motivate and people prefer soft postings.
- Lack of competition due to monopolistic market over a long period has led to complacent and noncompetitive work culture in the organization (supplemented with demand for oil and gas exceeding supply all through).
- Low productivity compared to international oil companies (IOCs) contract rigs.
- Lackluster reward and punishment policy

(e) The organization ranks high in terms of profitability, market capitalization, and so on. among global energy companies, but it ranks low in performance of HR parameters. A joint survey conducted by an internationally acclaimed consultancy firm and a respected business magazine revealed that this E&P company does not figure in the list of "Best 25 Employers in the Country."

(f) The low level of employee engagement was observed across the organization on various counts. For example, only 30% of employees are registered with the company intranet, even though 60% population in the company are designated as officers. This reflects low degree of employee engagement and motivation in the organization.

Discipline Related

(g) Discipline-related issues were largely overlooked by the administration/management even though nagging HR issues persisted in the organization. For example, no man-days were reportedly lost due to industrial relation (IR) or internal reasons during the past 3 years in this Asset. Even though, there were several agitations by the employees in the form of pen/tool down, work-to-rule, dharna, hunger strike, and so on, during this period. A section of the workers often disregarded their seniors flaunting their allegiance to the trade unions, as indiscipline perpetuated by the trade union members was mostly ignored. This had adverse effect on the work culture of the organization, which in turn affected operational performance.

All these portray a lackadaisical HR policy, inept HR administration, and leadership, which were partly responsible for deteriorating work culture, low motivation, and involvement of a large number of employees in the organization.

Food Habits and Health

This is an interesting aspect although not directly linked with performance of work but has bearing on the health of employees, especially those working in offshore and on-land rigs/fields on the long term.

(h) A normal man requires 2800 kcal in a day for performing moderately strenuous job and 3900 kcal for hard manual work. A study carried out by a reputed institute some time back revealed that the food provided daily at the company camps/drilling rigs ranges between 4346 and 4945 kcal. This has further increased to

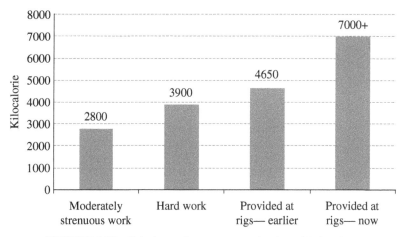

FIGURE 2.18 Calorie requirement versus food provided at rigs.

7000+ kcal at present, which is much higher than the requirement of the human body. This is graphically shown in Figure 2.18.

Considering the average age of crews working at drilling rigs is 42.2 years and 68% of the crews is aged above 40 years, such high-calorie food (beyond the requirement of the human body) provided daily to the aging crews may prove:

- Detrimental to health
- Genesis of physical illness
- Counterproductive in the long run

2.6.2 Suggestions for Improvement

Based on the findings and discussions in Section 2.6.1, the following suggestions are made to improve HR planning and policy, motivation level of employees, awareness on healthy food habits to rig crews, and so on, which have bearing on drilling performance.

HR Planning and Policy: Training and Development

(a) The crew members of own rigs may be sent on education tour to hired/private operator's rigs for familiarization with their working and work culture.

(b) Intensive training to employees, especially operation/field personnel may be imparted in the areas of:
- Adopting competitive work practices
- Changing mind-set
- Instilling confidence

Motivational Aspects

(c) In order to improve motivation of rig crew suitable awards may be instituted for "Best Drilling Rig of the Year (own)" that surpasses performance of the best contract rig.

(d) Inter-rig movement of crew may be encouraged and their posting may be rotated every 3–4 years. Such rotations would provide greater exposure and learning opportunities, motivation, and so on.

Food Habits and Health

(e) Considering the age and health of crew members at drilling rigs, it is suggested that:
  • Rig crews may be educated on health awareness and counseled on food habits and good health. A committee comprising representatives of collectives (trade unions/employee associations), drilling services, HR, and medical officer may be formed for this purpose.
  • Posters and charts may be displayed at the dining hall of rigs/camps, highlighting the following: (i) Nutritive values of different food/meals/dishes, (ii) calorie requirement, (iii) benefits of taking leafy vegetables and whole pulses, and (iv) effect of consuming animal fat, fried food, beverages, and so on.
  • Revise daily food chart of crew onboard through professional dietician. Provision of food may be made such that intake calorie does not exceed 3900.

2.7 TECHNICAL FACTORS

Technical factors include maintenance practices and planning, drilling practices and their effect on nonproductive drilling time, and drilling time slippage. This section contains two subsections, namely, (i) areas of concern and (ii) suggestions for improvement.

2.7.1 Areas of Concern

The maintenance planning and practices followed by own rig crew is not at par with that of contract rigs. There is considerable disparity, which is evident from the following fact:

(a) The average equipment downtime per rig per year is 77% higher in own rigs than hired rigs (own: 392 h, hired: 222 h; Fig. 2.19).

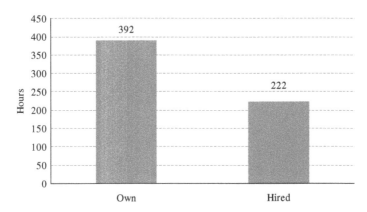

FIGURE 2.19   Average equipment downtime per rig per year (h).

It has been observed that the scheduled/preventive maintenance of critical rig equipment is often deferred in own rigs due to the nonavailability of spares and materials, which in the long run results in higher equipment downtime.

Drilling Practices
There is a marked difference in drilling practices followed by own rig crew and that of contract rig. Safe and sound drilling practices minimize drilling complications, time slippage, and enhance drilling efficiency. The analysis of phase-wise drilling, nonproductive drilling time in own and contract rigs reveals startling facts, which are discussed in the following paragraphs:

(b) The average drilling time slippage at each phase is higher in own rigs than that of hired rigs. Many a times, contract rigs complete jobs before the scheduled completion time. The comparison of phase-wise (30/26″, 20″, 13-3/8″, 9-5/8″, 7″, 5″) drilling time slippage for Year 1 and Year 2 is shown in Figures 2.20 and 2.21, respectively.

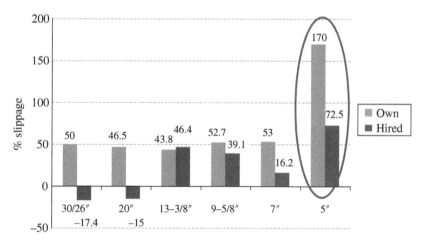

FIGURE 2.20   Comparison of phase-wise drilling time slippage (%), Year 1.

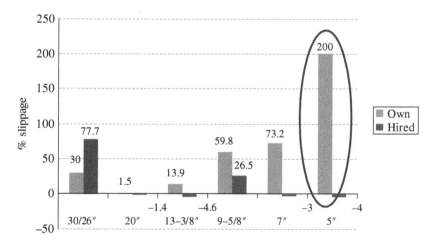

FIGURE 2.21   Comparison of phase-wise drilling time slippage (%), Year 2.

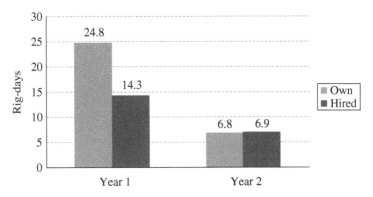

FIGURE 2.22 Average nonproductive time due to stuck-up, fishing, and sidetracking per rig per year.

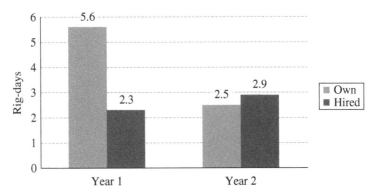

FIGURE 2.23 Average nonproductive time due to downhole complications per rig per year.

It may be noted that drilling time overrun below 7″ casing depth is alarmingly high in own rigs, 170% in Year 1 and 200% in Year 2, which requires attention.

(c) The average nonproductive time due to stuck-up, fishing, and sidetracking per rig per year is 73% higher in own rigs than that of hired rigs in Year 1 (own: 24.8, hired: 14.3 rig-days). However, there is marked improvement in Year 2, and nonproductive time on account of these in own and contract rigs is nearly the same (own: 6.8, hired: 6.9 rig-days). These are shown in Figure 2.22.

(d) The average nonproductive time due to downhole complications per rig per year is 143% higher in own rigs than that of hired rigs in Year 1 (own: 5.6, hired: 2.3 rig-days). But there is improvement in the performance of own rigs in Year 2 (own: 2.5, hired: 2.9 rig-days) and is shown in Figure 2.23.

Drilling practices are partly responsible for higher nonproductive drilling time in own rigs due to (i) stuck-up, fishing, sidetracking, and (ii) downhole complications.

2.7.2 Suggestions for Improvement

Based on the aforementioned findings and discussions in Section 2.7.1, the following suggestions are made for improvement in drilling practices, maintenance practices and planning, and so on, which would minimize drilling complications and nonproductive drilling time.

Drilling Practices

(a) Drilling through critical sections prone to complications (i.e., below 7" casing depth for the instant case) in own rigs may be carried out under the direct supervision of a senior drilling professional and may be closely monitored by higher level drilling experts.

(b) Periodic training may be imparted to drilling crew on downhole complications. Special training may be given for skilful operations of TDS, pile hammer, and so on.

(Further suggestions on safe drilling practices are discussed in Chapter 4.)

Maintenance Practices

(c) Greater emphasis may be given to maintenance audit of critical rig equipment, and enforce strict monitoring. Ensure compliance of maintenance guidelines and policy, especially for critical rig equipment.

(d) Awards such as "Best Maintained Drilling Rig of the Year" may be introduced at each Asset/Basin. The parameters for assessment may include, among others, rig performance, rig/equipment downtime, cost overrun, and the like.

(e) Introduce rate contract and operations and maintenance (O&M) contract to original equipment manufacturers (OEMs) for more category and number of materials and services. This would reduce the procurement lead time and ensure timely availability of better quality materials for maintenance and operations.

(f) Drilling rigs may be upgraded in a planned manner without affecting the planned activities, so that these are equipped with contemporary technology and achieve better performance standard.

(g) Emphasize on improvement of housekeeping in own rigs. This would help in ameliorating work environment, employee morale and productivity, reduce accident rate, and so on.

Planning

(h) *Inculcate cost consciousness*: Each rig unit may be encouraged to participate and plan for drawing internal baseline for drilling schedule, cost, and resources. They may compare their performance against this internal baseline and take corrective measures to meet the target. This will ensure greater engagement and participation of more employees.

Well cost estimates and actual well cost data may be circulated to all concerned to impress upon cost-effectiveness of drilling and related operations.

(i) It was observed that 240 rig-days costing USD 14 million (considering, average rig operating cost of owned and contract rigs combined at USD 58,500 per day×240 rig-days) were lost due to waiting on materials during the year in this Asset alone. In order to minimize the downtime on account of waiting on materials, it is suggested that:
  • Weekly/fortnightly material delivery schedule to drilling rigs may be prepared by drilling services indicating the date of delivery to rig. The schedule may be handed over to logistics department well in advance, so that they can plan for journey schedule optimizing capacity utilization of OSVs. This practice may be introduced and followed except in case of exigency.

- Accordingly, the logistics department may plan loading of OSVs optimizing floor space, weight, journey time, route, and so on. This would improve capacity utilization of OSVs, reduce number of OSV trips, and would lead to substantial savings.

(j) **Life cycle costing** may be introduced in the bidding practices, instead of lowest purchase price for procurement of energy-intensive equipment and materials, wherever feasible. The following parameters may be considered for life cycle costing:
  - Initial cost (purchase price, freight, installation, etc.)
  - Operation/maintenance cost
  - Energy cost
  - Disposal cost/salvage value
  - Life of equipment/item

(k) There is considerable scope for savings on account of optimum utilization and judicious deployment of offshore rigs for drilling and workover operations. It is observed that overcapacity rigs are often deployed for workover operations, which assumes significance in light of wide variations in rig rentals and operating cost for different types and capacities of offshore rigs.

2.8 BENEFITS AND SAVINGS POTENTIAL

All these measures suggested in Sections 2.5.2, 2.6.2, and 2.7.2 would help in containing time overrun of drilling operations, timely well completion, and improve the productivity of drilling operation with the *potential of savings around USD 60.5 million per year* in this offshore Asset. The details of which are elaborated in the following paragraphs:

(a) If drilling and well completion activities by own rigs are completed as per planned schedule, there is a scope of *savings of 1090 rig-days or USD 50 million* (considering average rig operating cost of own rigs at USD 46,000 per day × 1090 rig-days). This is a feasible and achievable target as the work program and drilling plan is prepared as per norm. It is observed that many wells are completed in time, and some wells (especially by hired rigs) are completed before the scheduled completion time.

(b) If the position of crane operator, welder, rigman, and roustabout are hired or outsourced, there *would be a saving of USD 2.25 million per year*. The details are shown in the following table.

	Crane Operator	Welder	Rigman	Roustabout
Existing strength at own rigs (10 nos.) ... [a]	49	26	120	223
Required for 10 rigs (if hired, as per deployment of contract rigs) ... [b]	40	20	80	200
Average annual salary at own rigs (USD thousand) ... [c]	9	9	5.25	9.25
Average annual salary at contract rigs (USD thousand) ... [d]	7	6	2.75	2.5
Annual savings (USD thousand) $[(a-b)c+b(c-d)]$	161	114	410	1563

(c) If the strength of crew at own rigs is kept at the level of contract rigs, there *would be a saving of USD 3.1 million per year* (considering the difference in strength of crew per rig (101–82), 19 nos. × 10 rigs × average annual salary per employee at USD 16,500).

(d) Advance weekly plan of material delivery schedule to drilling rigs would improve capacity utilization of OSVs in terms of floor space, weight, schedule of journey, and reduce the number of OSV trips. Considering all these efforts would reduce OSV cargo cost by a modest 10%, this *would save USD 3.5 million per year* (refer Section 3.5).

And 10% reduction of drilling rigs waiting on materials *would save USD 1.5 million per year* (refer to Section 3.5).

2.9 LIMITATIONS OF THE STUDY AND SCOPE FOR FURTHER WORK

This study deals with improving the productivity of drilling operations and focuses on operational productivity at a composite level taking into consideration drilling speed (cycle speed, commercial speed), well completion, drilling practices, maintenance practices, etc. It may be noted that productive and nonproductive rig time are intrinsic to these parameters, and drilling cost is directly influenced by these factors.

Operational productivity emphasizes on drilling speed and assumes importance due to high rig hiring and operating cost. But, it is often found that compromised drilling operation leads to unsafe drilling practices that may result in downhole complications or might damage the formation whose economic impact would be enormous. Therefore, further work can be done taking these aspects into consideration, that is, completion of wells in time with minimum damage to the formation and with high mechanical integrity to deliver "a healthy well at optimum cost."

Furthermore, the purpose of exploratory and development drilling goes beyond drilling speed, footage drilled per day, and drilling efficiency. The primary objectives of exploratory drilling are reserves accretion and collection of geological information and subsurface data. Similarly, the purpose of development drilling is to produce oil and gas. But, these aspects and their economic consequences have not been considered in detail in the current study. Further work can be done taking these parameters into consideration, which would broaden the scope of drilling productivity and add a new dimension to it.

2.10 CONCLUSION

The study reveals many interesting facts—performance of hired rigs is consistently and considerably better than the own rigs in terms of cycle speed, commercial speed, well completion, drilling time slippage, drilling practices, maintenance practices and planning, and so on. This is despite the fact that hired rigs are older than own rigs, contrary to the long-held belief of the company officials. The operational data for 2–3 years were analyzed, benchmarked against contract rig performance, and the factors for low performance of own rigs were identified. These were grouped in three categories, namely, human factors, organizational factors, and technical factors. The areas of concern and suggestions for improvement of these factors are summarized in the following paragraphs:

Areas of Concern

(i) *Human factors*: The aging and overstaffed crews posted at deep drilling rigs, archaic staffing pattern, work practices, and culture despite the advancement of technology, inappropriate officers to workers ratio at drilling rigs, officers with inadequate knowledge base and inappropriate skill sets posted at drilling rigs, and so on. have bearing on the productivity of drilling operations.

(ii) *Organizational factors*: Lackadaisical HR planning and policy, especially replacement of aging crews, entry-level induction, training and development of critical discipline and category employees, and career progression policy; inept HR administration—handling of discipline-related issues, low level of employee motivation, and engagement.

(iii) *Technical factors*: Drilling practices, maintenance practices, and planning in own rigs are not at par with the hired rigs leading to drilling complications and higher nonproductive drilling time. There is ample scope for reduction in nonproductive time due to stuck-up, fishing, sidetracking, downhole complications, equipment failure, and so on, in own rigs.

Suggestions for Improvement

(i) *Human factors and organizational factors*: Formulate and implement effective HR plan and policy for replacement of aging crews, induction at entry level, especially the critical and manual labor-intensive categories; improve competency level of rig crews through induction, special training, and posting of people with appropriate skill sets at drilling rigs.

(ii) Conduct special training on behavioral change, motivation and participation, and so on, across the organization; promote fair and transparent HR policy including career progression and rotation policy; effective leadership and HR administration to deal with discipline-related issues.

(iii) *Technical factors*: Improve drilling practices, maintenance practices, and planning in own rigs to minimize drilling time slippage and nonproductive drilling time. Special attention is needed for drilling through critical sections prone to complications (i.e., below 7″ casing depth for the instant case); conduct special training on downhole complications, TDS, pile hammer, and so on, to drilling and maintenance crew; ensure compliance of maintenance guidelines and schedule for critical rig equipment; rig upgrading; introduce rate contract, O&M contract to OEMs for more number of equipment, materials, and services to shorten procurement lead time and improve quality of maintenance; and promote cost consciousness in planning, operations, and maintenance activities.

(The detailed suggestions for improvement are listed in Sections 2.5.2, 2.6.2, and 2.7.2.)

The suggested measures would help in minimizing drilling time slippage, nonproductive drilling time, timely completion of wells, and optimize productivity of drilling operation with the *potential of savings of USD 60.5 million per annum* in this offshore Asset.

Chapter 3 deals with optimizing controllable rig time loss and minimize nonproductive drilling time. It follows a diagnostic approach to identify various elements of controllable rig time loss, finds out its root causes, and suggests corrective measures to optimize on-bottom drilling and completion time, which in turn would improve the productivity of drilling operation.

REVIEW EXERCISES

2.1 As a manager, how would you improve the productivity of drilling operations in your organization?

2.2 What are the factors that influence drilling efficiency?

2.3 What are the reasons for low performance of drilling operations in own rigs and areas of concern as per this study?

2.4 "Performance of hired rigs is consistently better than own rigs"—what are the lessons learnt that can be applied to own rigs as per this study?

2.5 How does human factor impact the productivity of drilling operation? How would you improve it?

2.6 What do you understand by organizational factors that influence performance of drilling operation? Suggest a plan for its improvement.

2.7 What are the technical factors that influence the performance of drilling operation? How would you improve them?

Advance

2.8 As a drilling expert, what would you suggest to minimize downhole complications?

2.9 What are doglegs and key seats? What are the reasons for their occurrence and remedial measures?

REFERENCES AND USEFUL LINKS

[1] Joint Association Survey, Drilling Costs, American Petroleum Institute (API), Washington, DC, 1976–2004.

[2] U.S. Department of Energy/Energy Information Administration (USDOE/EIA), Indexes and Estimates of Domestic Well Costs 1984 and 1985, USDOE/EIA-0347 (84–85), USDOE/EIA, Washington, DC, November 1985.

[3] Snead, M.C., The Economics of Deep Drilling in Oklahoma, working paper, Oklahoma State University, Stillwater, OK, 2005.

[4] Cochener, J., Quantifying Drilling Efficiency, U.S. Energy Information Administration, available at: http://www.eia.gov/workingpapers/pdf/drilling_efficiency.pdf (accessed on February 27, 2016).

[5] Chowdhury, S., Striking Oil, Science Reporter, Volume 31, Number 5, May 1994, pp. 9–12, 35.

[6] Rigzone, Offshore Rig Day Rates, April 2015, available at: http://www.rigzone.com/data/dayrates/ (accessed on February 27, 2016).

FURTHER READING

Devereux, S., Drilling Technology in Nontechnical Language, 2nd edition, PennWell Publishing Company, Tulsa, OK, 2012.

Hyne, N.J., Nontechnical Guide to Petroleum Geology, Exploration, Drilling, and Production, 3rd edition, PennWell Publishing Company, Tulsa, OK, 2012.

Joshi, S.D., Horizontal Well Technology, PennWell Publishing Company, Tulsa, OK, 1991.

Rabia, H., Oil Well Drilling Engineering Principal and Practice, Graham and Trotman Limited, London, 1985.

Short, J.A., Drilling: A Source Book on Oil and Gas Well Drilling from Exploration to Completion, Pennwell Publishing Co., Pennwell Books, Tulsa, OK, 1974.

Useful Links

Drilling Technology and Costs, available at: https://www1.eere.energy.gov/geothermal/pdfs/egs_chapter_6.pdf (accessed on February 27, 2016).

Osmundsen, P., Roll, K.H., and Tveterås, R., Exploration Drilling Productivity at the Norwegian Shelf, University of Stavanger, available at: http://www1.uis.no/ansatt/odegaard/uis_wps_econ_fin/uis_wps_2009_34_osmundsen.pdf (accessed on February 27, 2016).

Wikipedia, Oil Well, available at: http://en.wikipedia.org/wiki/Oil_well (accessed on February 27, 2016).

3

A DIAGNOSTIC APPROACH TO OPTIMIZE CONTROLLABLE RIG TIME LOSS

3.1 INTRODUCTION

In Chapter 2, we have discussed about the capital-intensive drilling operation, its spiraling cost, reasons for low performance of in-house drilling services, and measures for improving the productivity of the drilling operation. These are important issues for survival of many E&P companies who are hard pressed to contain rising drilling cost and improve drilling productivity. In this chapter, we would deal with controllable rig time loss, which is part of nonproductive drilling time and has considerable scope for improvement.

It is necessary to minimize nonproductive drilling time and improve on-bottom drilling and completion time, as drilling is a costly operation consuming large resources of an E&P Asset. Nonproductive drilling time consumes substantial amount of drilling time ranging from 10% to as much as 60% depending on geological complexities and difficulties faced during drilling. Nonproductive time (NPT) is the duration when the drilling operation has either ceased or has negligible penetration due to downhole complications, waiting, or other reasons (as specified later). It is expressed as percentage of drilling time and has both reducible and irreducible components. Drilling plan takes care of irreducible NPT, but if it extends beyond permissible or planned limit (which varies from company to company), it would delay well completion and subsequent drilling or project plan. Such delay on account of nonproductive drilling time would adversely impact the economics of capital-intensive drilling project and investment plan.

This study uses diagnostic approach to identify reducible NPTs and optimize controllable rig time loss, which would ensure higher on-bottom drilling and completion time, and improve the productivity of the drilling operation. Sometimes, reducible NPTs such as

Optimization and Business Improvement Studies in Upstream Oil and Gas Industry, First Edition.
Sanjib Chowdhury.
© 2016 John Wiley & Sons, Inc. Published 2016 by John Wiley & Sons, Inc.

waiting on materials (WoM) and tools, equipment failure, and so on, are camouflaged under technical reasons to avoid penalty, which is partly responsible for reportedly high downhole complications. This study was carried out in a large offshore Asset whose fields and reservoirs are spread over an aerial extent of more than 90,000 km². The Asset has both own and hired drilling rigs and offshore supply vessels (OSVs). Apart from that, it has in-house and contract services for cementing, logging, well services, and logistics.

The rudimentary of drilling operation and basic idea of drilling rig and cost were discussed in Chapter 2. In furtherance to that, some more related issues and activities, which are useful to the context of this study, are explained in the Section 3.2.

3.2   RIG TIME LOSS AND NONPRODUCTIVE DRILLING TIME

The components of rig time loss, nonproductive drilling time, well design and drilling plan, and so on, are elaborated in this section, which would help the readers to comprehend the current problem and its context.

The decision to drill a well, especially exploratory well, requires diligent teamwork by geologist, geophysicist, petrophysicist, drilling engineers, and others. The well design and drilling plan are prepared before the start of drilling and is normally followed till the completion of the well. It is a dynamic plan and might change depending on the borehole conditions encountered during drilling. A typical well plan contains well location, target depth, water depth (for offshore), expected reservoir pressure, presence of hydrocarbon (oil/gas) and H_2S, evaluation needs (i.e., mud log, well log, and core sample), phase-wise drilling plan, activity timeline, expected drilling problems, mitigation plan, and so on.

Drilling plan is prepared taking into account several factors, such as formation geology, pore pressure analysis, fracture gradient prediction, and so on, and contains casing design, cementing plan, drilling fluids plan, completion design, drill string design, drill bit program, identification of downhole complication and subsequent mitigation plan, well logging plan, well testing, coring plan, rig selection, and so on [1, 2].

The selection of rig is generally guided by the principle of matching rig capacity with well depth, well type, and well size; in addition, rig availability is an influential factor for rig deployment. It specifically depends on the hook load capacity, deck space, blowout preventer (BOP) specification, drilling hydraulics requirement, drilling fluid maintenance and delivery system, water depth (for offshore), Health, Safety and Environment (HSE) requirement, and special rig requirement (i.e., desert rig, heli rig, small footprint rig, batch or pad drilling rig, etc.). Daily rig rentals and rig mobilization cost are also influential factors for rig selection.

Rig cycle time is the duration between rig release (from the current well) to rig release of the next well. Rig time comprises time for drilling, workover, logging, coring, production testing, rig move, and so on. Drilling time may be divided into productive and NPT. Drilling, workover, and evaluation times are considered as productive rig time. Logging, coring, and production testing time are known as evaluation time. *The components of nonproductive drilling time are stuck pipe (drill string or tool), fishing, lost circulation, well control, wiper trips related to wellbore instability, pipe failure, equipment failure, downhole complications, WoM and tools, waiting on weather (WoW) (especially for offshore), waiting on decision (WoD), and so on.* NPT has both reducible and irreducible components. While some activities such as WoW, and so on, are unavoidable, many others can be avoided or controlled to a great extent. *This study tries to identify the root causes of this controllable rig time loss and suggests measures to optimize these.*

Geologic formation (i.e., hardness and composition of rock, formation pressure, temperature, behavior, etc.) and characteristics of well (i.e., well type, location and depth of target reservoir, etc.) have a casting influence on NPT. It is observed that NPT is generally higher in hard, abrasive, and heterogeneous formation, which is prone to drill string failures. It is also noted that deep reservoir is generally associated with high pressure and high temperature, hydrogen sulfide (H_2S) and carbon dioxide (CO_2) contamination, and prone to downhole complications [3]. The well trajectory, drilling fluid characteristics, and hydraulics have important role in mitigating downhole complications, and thereby minimize nonproductive drilling time. Drilling plan recognizes these complexities, and appropriate measures are required to optimize nonproductive drilling time.

To know more about oil and gas exploration and production activities, interested readers may refer to "References/Further Reading" listed at the end of this chapter.

3.3 OBJECTIVES

It is the endeavor of all E&P companies and drilling operators to optimize rig time loss and improve the drilling productivity, which is necessary for survival. This study focuses on controllable rig time loss with the following objectives:

(a) To optimize controllable rig time loss and improve on-bottom drilling and completion time

(b) To maximize rig time availability, reduce nonproductive drilling time, and improve the productivity of the drilling operation by ensuring more time for drilling

3.3.1 Methodology

It is important that E&P companies maintain operational database, such as well data, drilling data, mud data, and so on, which would be helpful for monitoring operations, decision making, operational planning, analysis, preparation of various reports, and so on. The offshore Asset where this study was conducted maintains a good operational database.

- The data for both productive and nonproductive drilling operations for 1 year were collected and studied in detail.
- Nonproductive drilling time was critically examined, and the major components of rig time losses that are reducible and can be controlled to some extent were identified.
- The study uses diagnostic approach to find out root causes of rig time loss, highlights shortcomings, and suggests remedial measures.

3.4 OBSERVATIONS

The major components of controllable rig time loss were identified as WoM, WoD, waiting on logging (WoL) tools, equipment repair downtime, and other shutdown.

The controllable downtime as percentage of rig time (also in hours) was found to be WoM—3.3% (5755 h), WoD—0.17% (304 h), WoL—1.51% (2525 h), equipment repair downtime—3.91% (6813 h), and other shutdown—5.27% (9181 h), which is shown in Figure 3.1.

The frequency or occurrences of shutdown in a year on account of these elements are observed as WoM—519 times (18% of total occurrence), WoD—37 times (1%), WoL—230 times (8%), equipment repair—1229 times (41%), and other downtime—966 times (32%), which are shown in Figure 3.2.

The cost of various components of controllable downtime was found to be WoM—USD 14 million, WoD—USD 0.6 million, WoL—USD 6.1 million, equipment repair downtime—USD 18.2 million, and other shutdown—USD 21.6 million per year and is shown in Figure 3.3.

A comparison of average downtime between own rigs and contract rigs operating in the same offshore field/Asset reveals the following:

(i) The average WoM is nearly same for both own and hired rigs (own—256 h, hired—245 h; Fig. 3.4). This is because materials are supplied to rigs and installations by OSVs, which are controlled by asset logistics group. Priorities are usually given to hired rigs because of its high idling cost.

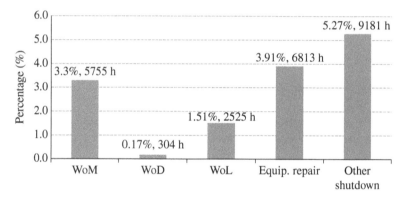

FIGURE 3.1 Elements of controllable downtime (h, as % of rig time).

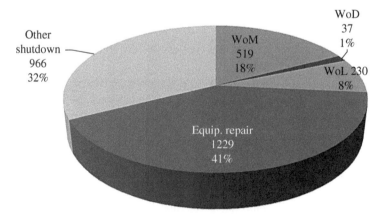

FIGURE 3.2 Occurrences of controllable downtime (number of times, % of occurrence).

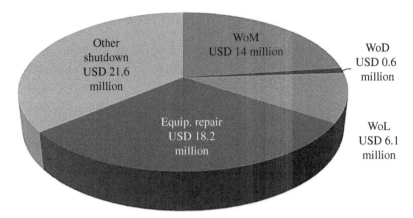

FIGURE 3.3   Components of controllable downtime: cost.

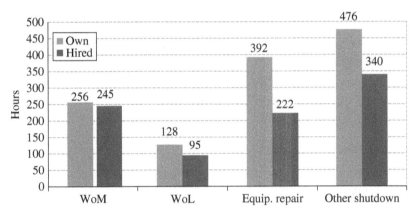

FIGURE 3.4   Average controllable downtime per rig per year.

(ii) The average WoL or unit in own rigs is 35% higher than that of hired rigs (own—128 h, hired—95 h; Fig. 3.4).

(iii) The average rig downtime due to equipment repair is 77% higher in own rigs than that of hired rigs (own—392 h, hired—222 h; Fig. 3.4). This was mentioned in Chapter 2.

(iv) Similarly, the average rig downtime due to other shutdown is 40% higher in own rigs than that of hired rigs (own—476 h, hired—340 h; Fig. 3.4).

3.4.1   Summary of Observations

The components of rig time loss during Year N in this offshore Asset are summarized in the following paragraph:

- Controllable rig time loss: 14.2% of rig time (24,578 h)
- Frequency or occurrence in a year: 2981 times
- Costs: USD 60.5 million/year
- Savings potential: USD 34.5 million/year

The controllable rig time loss during Year N comprises the following:

- WoM: occur 519 times, consume 3.3% of rig time, and cost USD 14 million.
- WoD: occur 37 times, consume 0.17% of rig time, and cost USD 0.6 million.
- WoL: occur 230 times, consume 1.51% of rig time, and cost USD 6.1 million.
- Equipment repair downtime: occur 1229 times, consume 3.91% of rig time, and cost USD 18.2 million.
- Other shutdown: occur 966 times, consume 5.27% of rig time, and cost USD 21.6 million.

3.5 WAITING ON MATERIALS

This section deals with WoM, which is a major component of controllable rig time loss. The root cause analysis reveals that rigs remain idle due to the following:

- WoM: **3.3%** of rig time (5755 h)
- Costing: USD 14 million
- Number of occurrences (frequency) in a year: 519 times
- Maximum waiting duration for a single occasion: 192 h (own rig 1 for mud chemicals).

3.5.1 Observations and Areas of Concern

The components of WoM along with duration (in hours and percentage of WoM), its causes, and subcomponents are summarized as follows (see Figure 3.5):

- Bulk (drill water, pot water, barite, cement, fuel, etc.): 2046 h (36%)
- Chemicals (mud chemicals, NaCl, $CaCl_2$, and mud/brine preparation): 1055 h (19%)
- Material (casing, well head, generator, equipment, spares, etc.): 1106 h (19%)

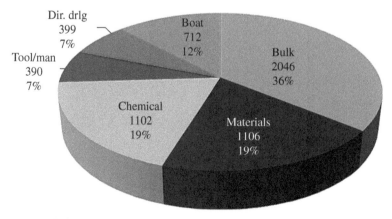

FIGURE 3.5   Elements of waiting on materials (h, % WoM).

- Directional drilling tools and men: 399 h (7%)
- Boat (towboat and AHSV): 712 h (12%)
- Other tools and men (fishing/DST/milling tool, fire, wire line, well head party, experts, diver, and crane operator): 390 h (7%).

WoM during different operations, namely, drilling, production testing, workover, and rig move as percentage of respective operation time (also in hours) is shown in Figure 3.6.

It reveals that WoM during rig move is relatively high (9.4%) and requires attention. This is mainly due to waiting for towboat, anchor handling boat, dive vessel, diver, and so on.

3.5.2 Duration of Waiting versus Frequency of Occurrence

Seventy-one percent (71%) of WoM are of duration 12 h and more, and the frequency of such occurrence is 35%. Merely 4.5% of WoM are less than 4 h duration, but the frequency of such occurrences is 25%. These are shown in Figure 3.7.

3.5.3 Availability of OSVs

OSV is the lifeline of offshore operations as the supply of materials to rigs and installations is largely dependent on it. Therefore, its availability is of paramount importance to an offshore Asset. This Asset has over 30 numbers of own OSVs and 23 hired OSVs. The following has been observed:

(i) The utilization of own OSVs is only 41.7%; in other words, out of 31 own OSVs, only 13 OSVs are effectively operative round the year.
(ii) In contrast, the utilization of hired OSVs is 93.2%, that is, 21 OSVs are operative round the year against available 22.5 OSVs (fraction indicates one or more OSVs were available for part of the year).

These are graphically shown in Figures 3.8 and 3.9.

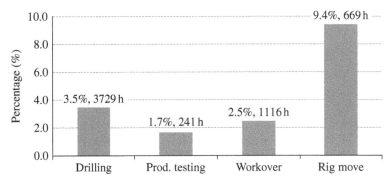

FIGURE 3.6 WoM during different operations (h, as % of respective operation time).

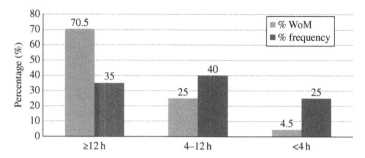

FIGURE 3.7   Waiting on materials: duration and frequency.

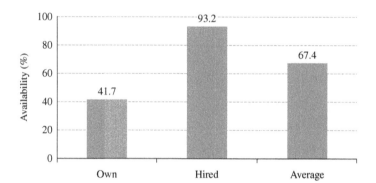

FIGURE 3.8   Utilization of offshore supply vessels (%).

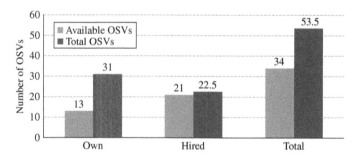

FIGURE 3.9   Availability of offshore supply vessels (number).

3.5.4   Suggestions for Improvement

Based on the aforementioned findings in Sections 3.5.1, 3.5.2 and 3.5.3, the following suggestions are made to minimize WoM:

(a) The coordination among indenting department, materials management, and logistics department at Asset base office and supply base must be improved with identified first person responsible (FPR).

(b) It has been observed that inadequate number of pallets and containers affect supply of chemicals. Therefore, provision for adequate number of pallets and containers can be made for facilitating the supply of chemicals.

(c) The concerned drilling in-charge (DIC) shall raise indent well in advance, so that materials arrive at rigs at least 1 day before its use.

(d) Improve utilization of water maker and reduce dependence on supply by OSV, thereby minimizing waiting on account of this.

(e) During Year N, the utilization or availability of own OSVs was only 41.7% compared to 93.2% that of hired OSVs. The main reason for such low utilization is the lack of maintenance and accountability. Immediate improvement in these areas is called for. The availability of own OSVs, if improved by 20–25%, then 6–8 more OSVs will be available for operation, which in turn, will reduce waiting on material.

(f) Waiting on towboat, anchor handling boat, and so on, during rig move is relatively high (9.4%). Since rig move plan is known in advance, rig move cell/team and logistic department must jointly prepare action plan and synchronize the movement of AHT boats and rigs.

(g) Introduce long-term contract with staggered delivery or "rate contract" for more frequently used materials. It will ensure better availability and will ease material movement planning.

(h) Adequate number of critical and frequently used spares may be kept at rigs to minimize waiting on account of these materials.

3.5.5 Savings Potential

If the aforementioned suggestions in Section 3.5.4 are followed, it is possible to reduce rig time loss on account of WoM from the current level of 3.3% to 0.5% (which is allowance for unforeseen and operational complications), and thereby savings of USD 12 million per year can be achieved.

3.6 WAITING ON DECISIONS

This section discusses WoD, which is an undesirable element of rig time loss. The analysis discloses that rigs remain idle due to the following:

- WoD: **0.17%** of rig time (304 h)
- Costing: USD 0.6 million
- Number of occurrences (frequency) in a year: 37 times
- Maximum duration of waiting for a single case: 120 h (own rig 2 for new location)

The reasons for WoDs are as follows:

- Waiting for location
- Waiting for program from G&G
- Waiting for log instruction

3.6.1 Observations and Areas of Concern

WoD during different operations, namely, drilling, production testing, workover, and rig move is shown in hours and as percentage of the total WoD in Figure 3.10.

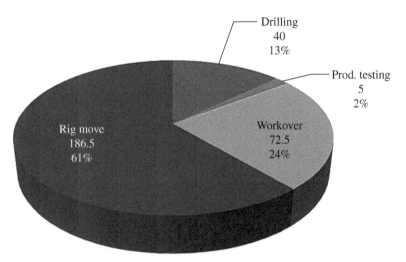

FIGURE 3.10 Waiting on decision during different operations (h, % WoD).

WoD is mainly found during rig move operation (186.5 h, 61%) for want of new location, followed by waiting for instruction and program from the base during workover operation (72.5 h, 24%), drilling operations (40 h, 13%), and production testing (5 h, 2%).

Waiting for location in one case took 120 out of 186.5 h (i.e., 64%) due to WoD, and it occurred during rig move operation. The reasons for such delay may be investigated and corrective measures can be taken accordingly.

3.6.2 Suggestions for Improvement

The following suggestions would minimize WoD:

(a) Rig shutdown for WoD is an example of management system failure and stem-ming from the lack of accountability. Rig shutdown due to WoD shall invariably be reported in daily/monthly "exception report" and submitted to the concerned director.

(b) As far as possible, persons on duty at drilling rigs may be empowered to take operational decision in their respective areas, which is found lacking.

(c) The FPR for decision such as well program, log instruction, next location, and so on, may be clearly identified and circulated to all concerned.

3.6.3 Savings Potential

The aforementioned suggestions in Section 3.6.2 will drastically reduce rig time loss on account of WoD, and thereby bring a saving of around USD 0.5 million per year.

3.7 WAITING ON LOGGING

This section covers WoL, which is a part of controllable rig time loss. The root cause analysis reveals that rigs remain idle due to the following:

- WoL and problem: **1.51%** of rig time (2525 h)
- Costing: USD 6.1 million
- Number of occurrences (frequency) in a year: 230 times
- Maximum waiting duration for a single case: 408 h (own rig 3 for logging unit)

3.7.1 Observations and Areas of Concern

The main causes of shutdown under this category are waiting for logging unit/tool/man—1680 h (67% of WoL), logging tool problem—583 h (23%), and logging tool stuck-up/fishing—262 h (10%) and is shown in Figure 3.11.

WoL during different operation, namely, drilling, production testing, workover, and rig move as percentage of respective operation time as well as in hours is graphically shown in Figure 3.12.

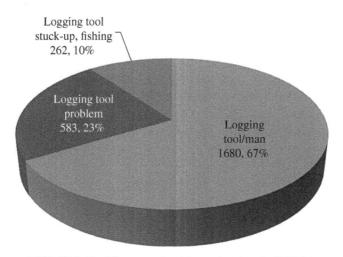

FIGURE 3.11 Elements of waiting on logging (h, % WoL).

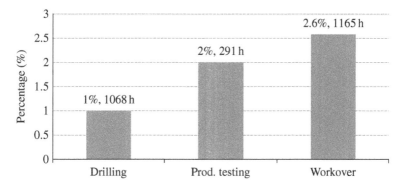

FIGURE 3.12 Waiting on logging during different operations (h, as % of respective operation time).

WoL as percentage of respective operation time is maximum during workover operation—2.6% (1165 h), followed by production testing—2% (291 h), and drilling—1% (1068 h).

3.7.2   Duration of Waiting versus Frequency of Occurrence

Sixty-three percent (63%) of WoL are of duration 12 h and more, and the frequency of such occurrence is 26%. Merely 4% of WoL are less than 4 h duration, but the frequency of such occurrences is 23%, *which is due to operational reasons* (Fig. 3.13).

3.7.3   Inferences and Suggestions for Improvement

Based on the aforementioned findings in Sections 3.7.1 and 3.7.2, the following suggestions are made to minimize WoL:

(a) The coordination among logistics department, asset monitoring group, and contractor needs further improvement.

(b) Logging engineer or expert (contractor), if manifested for a job on a particular date, priority may be assigned for sending the person, so that rig does not wait on account of this. (It is mentioned that helicopter services for ferrying people to various rigs/installations everyday have limited capacity and require prior manifestation and schedule.)

(c) A monthly exception report of rig time loss on account of WoL may be sent to the concerned director jointly by head of logging services and head of drilling services for drilling rigs, and jointly with head well services for workover rigs. Due to lesser contractual value, logging operation at times lacks attention, but in reality WoL, and so on, is a costly affair as waiting/shutdown cost of offshore rig is enormous.

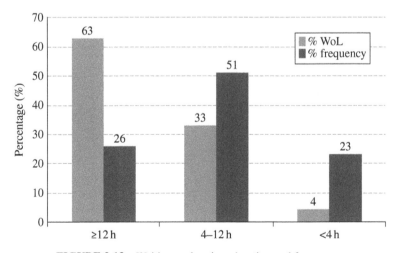

FIGURE 3.13   Waiting on logging: duration and frequency.

(d) While reporting stuck-up jobs during logging operation, special tools and accessories that are required but not available at the site/rig may be intimated to monitoring group/contractor to avoid delay in mobilizing tools.

(e) WoL during workover operations (2.6%) and production testing (2%) as respective operation time is on higher side (Fig. 3.12), although these operations are known well in advance. It indicates deficiency in planning, which needs to be improved upon. Also, the fact that 63% WoL are of duration 12 h and more (Fig. 3.13) suggests the need for improvement in planning.

3.7.4 Savings Potential

The aforementioned suggestions in Section 3.7.3 will cut down WoL and improve rig availability. With better coordination of activities, preplanning, and close supervision, 75% of waiting on account of logging unit/tool/man can be avoided, which would save approximately USD 3 million per year.

3.8 EQUIPMENT REPAIR DOWNTIME

This section discusses equipment repair downtime, which consumes considerable amount of rig time loss. The root cause analysis unearths that rigs remain idle due to the following:

- Equipment repair downtime: **3.91%** of rig time (6813 h)
- Costing: USD 18.2 million
- Number of occurrences (frequency) in a year: 1229 times
- Maximum waiting duration for a single occasion: 288 h (rig D2 for BOP repair)

3.8.1 Observations and Areas of Concern

The components of equipment repair downtime during Year N include repairing of the following:

- Mud pump: 1342 h (20% of equipment repair downtime)
- Top drive: 1320 h (19%)
- BOP: 1216 h (18%)
- Draw work: 445 h (7%)
- Engine/electrical system: 498 h (7%)
- Pile hammer: 285 h (4%)
- Other rig equipment/tool: 1707 h (25%)

This is graphically shown in Figure 3.14.

Equipment repair downtime during different operations, namely, drilling, production testing, workover, and rig move as percentage of respective operation time and in hours is shown in Figure 3.15.

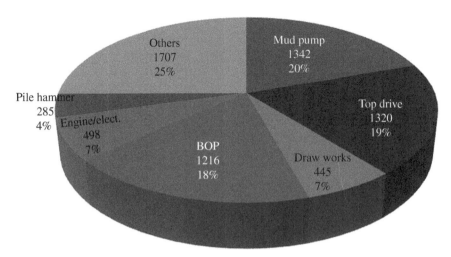

FIGURE 3.14   Elements of equipment repair downtime (h, % of equip. repair).

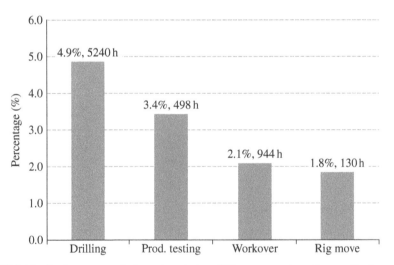

FIGURE 3.15   Equipment repair downtime during different operations (h, as a % of respective operation time).

Equipment repair downtime as percentage of respective operation time is highest during drilling operation—4.9% (5240 h), followed by production testing—3.4% (498 h), workover—2.1% (944 h), and rig move—1.8% (130 h).

3.8.2   Duration of Waiting versus Frequency of Occurrence

Fifty-seven percent (57%) of equipment repair downtime are of duration 12 h and more, and frequency of such occurrence is 11%. Only 19% of equipment repair downtime are less than 4 h duration, but frequency of such occurrences is 69%. This is depicted in Figure 3.16.

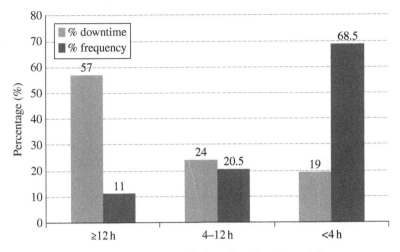

FIGURE 3.16 Equipment repair downtime: duration and frequency.

FIGURE 3.17 Equipment downtime per rig per year (h): comparison.

As per contract, allowable equipment downtime per rig per year is 432 h. This appears to be on higher side when compared to the actual downtime of hired rigs—222 h and that of own rigs—392 h (Fig. 3.17). In this regard, comparison may be made with international benchmark.

3.8.3 Inferences and Suggestions for Improvement

Based on the aforementioned findings in Sections 3.8.1 and 3.8.2, the following suggestions would minimize equipment repair downtime:

(a) Dry-docking schedule is not strictly followed due to operational reasons and necessity. Ensure dry-docking of all jack-up rigs as per planned schedule.

(b) There is a need to introduce skill improvement program for operation and maintenance of pile hammer and top drive system, and encourage introduction of long-term rate contract for repair. Sixty-five (65) rig-days were lost on account of these two equipment, of which, at least 48 rig-days costing USD 3.1 million could have been saved.

(c) Repair of mud pump modules may be undertaken only through reputed companies.

(d) The average equipment downtime per own rig is found to be 77% higher than that of contract rig. The equipment downtime of own rigs may be planned and brought down to the level of hired rigs. With this benchmarking, 71 rig-days costing about USD 4.5 million can be saved.

(e) Eleven percent (11%) occurrences of equipment shutdown are equal to or more than (≥) 12 h duration and account for 57% of equipment repair downtime with average downtime per occurrence being 28.8 h (Fig. 3.16). The frequency of equipment shutdown less than 4 h is due to normal operational reasons. The higher percentage of occurrences in this category indicates proper health of the equipment. On the contrary, the higher percentage of occurrences of equipment downtime more than 12 h indicates poor preventive maintenance. Since most of the drilling rigs are having redundant equipment, the average downtime in this category can be brought down considerably by using effective maintenance practices and policies.

(f) Equipment repair downtime of 3.4% during production testing (Fig. 3.15) is on the higher side considering rig equipment operating hours are less compared to drilling operations. This indicates complacency during the testing phase, which can be minimized by close supervision and monitoring.

(g) Equipment shutdown 4.9% during drilling operation (Fig. 3.15) is on the higher side and indicates a lack of supervision and maintenance. This can be improved upon and the equipment downtime during drilling operation may be limited to 3–3.5%.

(h) With alert supervision system and focused attention, downtime under this category can be brought down maximum up to 6 h per occurrence, which is usually allowed by E&P companies to drilling contractors. This would lead to a saving of approximately USD 8.25 million per year.

But, this cannot be achieved overnight and requires better planning and improved maintenance system. In order to achieve this, if at the first stage, downtime per occurrence is limited up to 12 h, it would bring about savings of USD 6 million per year.

(i) Emphasize on technology upgrade, wherever feasible. This would improve efficiency and reduce cost.

(j) Replace aging rig crews with younger and skilled workforce, which would improve rig productivity.

3.8.4 Savings Potential

The aforementioned suggestions in Section 3.8.3 have potential of savings of USD 12.5 million per year at a conservative estimate.

3.9 OTHER SHUTDOWN

This section deals with other shutdown, which consumes significant amount of rig time loss. The root cause analysis shows that rigs remain idle due to the following:

- Other shutdown: **5.27%** of rig time (9181 h)
- Costing: USD 21.6 million
- Number of occurrences (frequency) in a year: 966 times
- Maximum waiting duration for a single case: 168 h (own rig 4 for unable to release legs)

3.9.1 Observations and Areas of Concern

The causes and components of other shutdown are downhole complications (DH compl.), measurement while drilling (MWD), subsea, and others.

Downhole complications account for 65% (5932 h) of other shutdown time, followed by MWD 13% (1235 h), subsea 6% (568 h), and others 16% (1446 h), which is shown in Figure 3.18.

Other shutdown during different operations, namely, drilling, production testing, workover, and rig move as percentage of respective operations time and in hours is shown in Figure 3.19.

Other shutdown as percentage of respective operation time is highest during workover operation—7.5% (3386 h), followed by rig move—5.6% (400 h), drilling—4.6% (4926 h), and production testing—3.2% (469 h).

3.9.2 Duration of Waiting versus Frequency of Occurrence

Sixty-eight percent (68%) of other shutdown time are of duration 12 h and more, and frequency of such occurrences is 26%. Only 8% of other shutdown time are less than 4 h duration, but frequency of such occurrences is 42%. This is depicted in Figure 3.20.

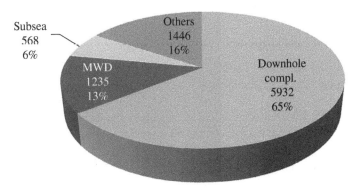

FIGURE 3.18 Elements of other shutdown (h, % of other shutdown).

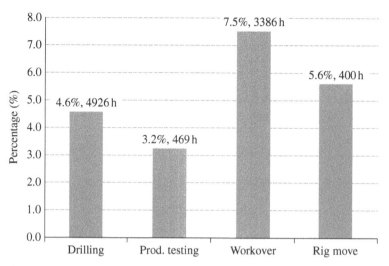

FIGURE 3.19 Other shutdown during different operations (h, as a % of respective operation time).

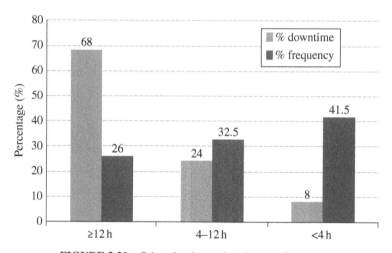

FIGURE 3.20 Other shutdown: duration and frequency.

3.9.3 Inferences and Suggestions for Improvement

Based on the aforementioned findings in Sections 3.9.1 and 3.9.2, the following suggestions would minimize other shutdown:

(a) Emphasize and embark on improvement in maintenance practices, planning, procurement of spares, and equipment (especially MWD).

(b) Introduce periodic training to drilling engineers on "downhole complications" to improve skill and workmanship.

(c) Other shutdown per own rig is found to be 40% higher than that of hired rig (Fig. 3.4). This can be brought down to the level of hired rig with proper planning.

(d) The higher percentage of other shutdown time during workover operations (7.5%) indicates poor planning and execution of workover jobs (Fig. 3.19), which can be improved upon by proper planning and supervision.

(e) Sixty-eight percent of other shutdown time is of duration 12 h and more, the majority of which is due to downhole complications (Fig. 3.20). It suggests that either the operations/procedures are not optimized or drilling is carried out through difficult formations. It calls for improvement in planning and drilling practices.

3.9.4 Savings Potential

Following the aforementioned suggestions in Section 3.9.3 and considering a modest reduction of 30% downtime on account of other shutdown would lead to a saving of USD 6.5 million per year.

3.10 WAITING ON WEATHER

This section covers rig time loss due to WoW. WoW is *not a controllable downtime* and has *not been considered under controllable rig time loss*. However, its effect can be controlled to some extent, therefore, WoW has been analyzed for finding out possible ways to minimize its effect on rig time loss.

The root cause analysis discloses that rigs remain idle due to the following:

- WoW: **1.8%** of rig time (3078 h)
- Costing: USD 15 million
- Number of occurrences (frequency) in a year: 217 times
- Maximum waiting duration for a single case: 696 h (rig J3 for lowering BOP)

3.10.1 Observations and Areas of Concern

Monsoon period which is only 4 months duration accounts for 55% of WoW (1693 h), while nonmonsoon period (8 months) accounts for the remaining 45%, 1385 h (Fig. 3.21).

WoW during different operations, namely, drilling, production testing, workover, and rig move as percentage of respective operations time and in hours is shown in Figure 3.22.

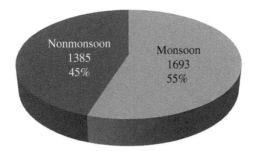

FIGURE 3.21   Waiting on weather (h, % WoW).

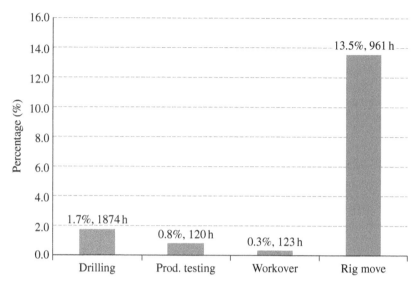

FIGURE 3.22   Waiting on weather during different operations (h, as a% of respective operation time).

WoW as percentage of respective operation time is highest during rig move operation—13.5% (961 h), followed by drilling—1.7% (1874 h), production testing—0.8% (120 h), and workover—0.3% (123 h). Apart from bad weather days, rig move as well as boat handling jobs is also affected due to rough sea condition.

3.10.2   Duration of Waiting versus Frequency of Occurrence

Seventy-three percent (73%) of WoW are of duration 12 h and more, and the frequency of such occurrences is 27%. Only 4% of WoW are less than 4 h duration, but the frequency of such occurrences is 32% (Fig. 3.23).

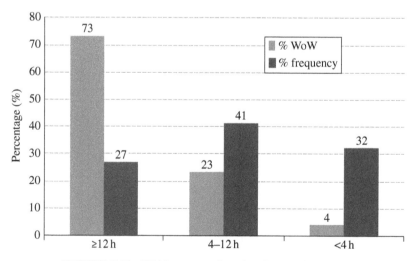

FIGURE 3.23   Waiting on weather: duration and frequency.

3.10.3 Suggestions for Improvement

The following suggestions have potential to minimize WoW:

(a) The effect of adverse weather condition in monsoon period can be reduced with induction of powerful boats and equipment. High-powered dynamic positioning (DP) supply vessels may be acquired to cater the need of number of rigs and platforms based on economic consideration.

(b) Rig time loss due to WoW in nonmonsoon period can be drastically reduced with better planning and coordination.

3.10.4 Savings Potential

Following the aforementioned suggestions in Section 3.10.3 and considering a modest reduction of 20% of downtime on account of WoW would bring about a savings of USD 3 million per year. As WoW has not been considered as controllable rig time, this savings has not been included in the savings potential for controllable rig time loss.

3.11 CONCLUSION AND SCOPE FOR IMPROVEMENT

The study uses diagnostic approach to identify major components of controllable rig time loss, which accounts for 14.2% of available rig-days costing around USD 60.5 million per year. There is ample scope for reduction in rig time loss, which has been identified and grouped as WoM (consumes 3.3% of rig time and costs USD 14 million), WoD (consumes 0.17% of rig time and costs USD 0.6 million), WoL (consumes 1.51% of rig time and costs USD 6.1 million), equipment repair downtime (consumes 3.91% of rig time and costs USD 18.2 million), and other shutdown (consumes 5.27% of rig time and costs USD 21.6 million).

The root causes of rig time loss for each of this group have been identified and remedial measures have been suggested, which are summarized in the following paragraphs:

(i) Synchronize and strengthen coordination among various groups/departments for different activities and identify responsible persons accordingly, for example, supply of materials, tools, and equipment to rigs/installations.

(ii) Improve utilization and availability of own OSVs.

(iii) Introduce specific skill improvement training for identified groups, for example, operation and maintenance of pile hammer, top drive system, mud pump, and other critical equipment.

(iv) Introduce long-term rate contract for specific jobs, critical and frequently used materials, spares, and so on.

(v) Ensure that some specific jobs are done only through reputed companies or contractors, for example, mud pump modules, repair and maintenance of critical equipment, and so on.

(vi) Shutdown due to WoD needs to be highlighted and reported as "exception report." FPR for respective decisions may be clearly identified and accountability be fixed. DIC and others on duty may be further empowered for taking operational decision in their respective areas.

(vii) Equipment repair downtime, especially for own rigs, needs to be brought down to the level of international E&P companies. Ensure planned dry-docking of jack-up rigs, technology upgrade, scheduled/preventative maintenance, and so on.

(viii) Special attention may be given to specific jobs like logging, downhole complications, and so on. Ensure qualitative improvement in maintenance practices, planning, procurement of spares and equipment, and supply of materials to rigs and installations.

(The detailed suggestions for improvement are presented in Sections 3.5.4, 3.6.2, 3.7.3, 3.8.3, 3.9.3, and 3.10.3.)

There would be considerable improvement in rig time availability if the aforementioned suggestions are followed in earnest. These would help in bringing down the controllable rig time loss from the current 14.2% to competitive level of 6%, which would entail a savings of USD 34.5 million per year (excluding savings on WoW, which has not been considered as controllable rig time). This is the savings potential in one offshore Asset and would multiply if replicated in other Assets and Basins of the company.

3.11.1 Scope for Improvement

As easy hydrocarbon reserves are becoming scarce, the E&P companies are venturing into difficult and marginal areas in search of oil and gas. Thus, the focus is shifting to deepwater exploration and in environmentally sensitive areas (e.g., the arctic region), super deep wells, long extended reach wells, difficult horizontal and multilateral wells targeting multiple reservoir from single location, EOR wells, and reservoir pressure maintenance (water flooding, chemical injection, etc.). These possess great technical and operational challenges. Therefore, rig selection, well planning, parameter optimization, technology requirement, and so on, are important factors and integral part of operation to achieve maximum productive rig time utilization and minimize NPT related to downhole complications.

Optimization is a continuous task requiring close monitoring and constant follow-up actions. Learning curve and subsequent improvements are usually gradual and time-consuming. Therefore, reducible NPTs are to be closely monitored and evaluated in subsequent years, so that the objectives or targets are met. Reducible NPT can be considerably brought down by establishing real-time drilling monitoring center at the base office that will operate and monitor round the clock. This center may be manned by a team of subject specialists with peer consultation and application software support, especially for difficult and high-value wells. The group can be empowered to take instant decision. Such type of dynamic and group decisions on real-time basis shall go a long way in reducing controllable NPT and improve drilling and completion efficiency. Further improvement in these areas may be initiated with the aim of achieving the global/IOC (international oil company) benchmark for reducible rig time losses.

Chapter 4 focuses on some G&G activities, which are not closely monitored, as these are associated with uncertainty and some degree of subjectivity. Nonetheless, there is much scope for improvement in these areas. It tries to improve the effectiveness of these G&G activities and optimize G&G strategies for deepwater oil and gas exploration.

REVIEW EXERCISES

3.1 What are the main components of controllable rig time loss? As a manager, how would you optimize controllable rig time loss?

3.2 What are the main components of nonproductive drilling time? What are your suggestions to minimize them?

3.3 What are the root causes of (a) waiting on materials (b) waiting on decision, and (c) waiting on logging? How would you minimize these?

3.4 What are the main components of (a) equipment repair downtime and (b) other shutdown as per this study? How would you minimize these?

Advance

3.5 What are the different types of pipe sticking and their reasons? What preventive measures would you take to avoid them?

3.6 What are the main reasons of pipe failures? As a drilling specialist, what would you suggest as remedial measures?

REFERENCES

[1] Adams, N.J., Drilling Engineering: A Complete Well Planning Approach, PennWell Publishing Company, Tulsa, OK, 1985.

[2] Marbun, B., Aristya, R., Pinem, R.H., Ramli, S.B., and Gadi, K.B., Evaluation of Non-Productive Time of Geothermal Drilling Operations—Case Study in Indonesia. In 38th Workshop on Geothermal Reservoir Engineering, Stanford University, Stanford, CA, February 24–26, 2013, available at: https://pangea.stanford.edu/ERE/pdf/IGAstandard/SGW/2013/Marbun4.pdf (accessed on March 2, 2016).

[3] Mark, J.K., A Survey of Drilling Cost and Complexity Estimation Models, International Journal of Petroleum Science and Technology, Volume 1, Number 1, 2007, pp. 1–22, available at: http://www.ripublication.com/ijpst.htm (accessed on March 2, 2016).

FURTHER READING

Azar, J.J. and Samuel, G.R., Drilling Engineering, PennWell Publishing Company, Tulsa, OK, 2007.

Messanger, J.U., Lost Circulation, PennWell Publishing Company, Tulsa, OK, 1981.

Useful Links

Drilling Performance Benchmark Matrix, available at: https://www.dodsondatasystems.com/Public/Files/Benchmark%20Metrics%20Public%20Web%20v1.pdf (accessed on March 2, 2016).

Eliminating Non-Productive Time Associated with Drilling Trouble Zones, available at: http://www.successful-energy.com/wp-content/uploads/2011/01/2009_OTC_20220_Eliminating_Non-Productive_Time.pdf (accessed on March 2, 2016).

Rig NPT: The Ugly Truth, available at: http://www.drillingcontractor.org/rig-npt-the-ugly-truth-6795 (accessed on March 2, 2016).

4

OPTIMIZING G&G STRATEGIES FOR DEEPWATER OIL AND GAS EXPLORATION

4.1 INTRODUCTION

In earlier chapters, we have dealt with optimizing controllable rig time losses and improving the productivity of drilling operations. This chapter focuses on optimizing some important geological and geophysical (G&G) activities and strategies for deepwater exploration.

As geologically and geographically easy reserves are fast depleting with the ever-increasing demand for oil and gas in this energy-hungry world, the focus has shifted to deepwater exploration, which is costly, risky, and technologically challenging. The cost and uncertainties involved in deepwater exploration are so high that major oil companies prefer joint venture operations for sharing cost and minimizing risks. E&P companies across the world are on continuous endeavor to contain the cost of deepwater exploration through various means including maximum utilization of available resources and exploration of all known and unknown opportunities.

Most of the E&P activities such as seismic survey, drilling, oil and gas production, workover jobs, and so on, are easily measurable and mostly visible. These activities are continuously monitored by the senior management and are much in attention. But some G&G activities are not as visible as other E&P activities, and their outcomes are not readily measurable. The performance of these G&G activities and strategies is generally not questioned under the wrap of subjectivity and uncertainties. This study aims to identify, assess, and improve the efficacy of some of these G&G activities, which are mentioned in the following paragraphs:

(i) Rig remains idle during evaluation (i.e., conventional logging and production testing (PT)), which runs for several days depending on the subsurface prospects and interest. Deepwater exploration is exorbitantly expensive—rig rental charges

Optimization and Business Improvement Studies in Upstream Oil and Gas Industry, First Edition.
Sanjib Chowdhury.
© 2016 John Wiley & Sons, Inc. Published 2016 by John Wiley & Sons, Inc.

are around USD 500,000 per day, and rig operating cost is even higher; therefore, it is necessary to optimize evaluation time and maximize productive drilling time.

(ii) Deepwater rigs are generally equipped with logging while drilling (LWD), which provides real-time data of subsurface properties. It enables drilling engineers and G&G professionals for real-time control, decision making, and avoidance of downhole complications. Much of the information received from LWD is also obtained through conventional logging later, as G&G experts consider both LWD and wireline logging necessary for various reasons. These have been discussed in detail in Section 4.4.1. The current study respects technical prudence of G&G experts and necessity of both LWD and wireline logging but considers there is scope for optimizing evaluation time in deepwater exploration with the opportunity of substantial saving.

(iii) The operating cost of own deepwater rig is much less than that of contract rig due to various reasons, such as type, class, and capacity of rig, age of rig, depreciation, and so on. The own deepwater rig under current study is capable of drilling up to 3000 ft water depth; beyond that contract rig is deployed. Due to high cost of LWD and low operating cost of own rigs that are operating in lesser water depth (1400–3000 ft), *the use of LWD in own rig may be governed by techno-economic and risk–reward evaluation.* These have been examined.

(iv) The geological predictions of hydrocarbon (HC) show depend on various parameters that are associated with high degree of uncertainty, low level of accuracy, and limitation in assessing or obtaining some parameter values. Therefore, geological predictions in unexplored area or unknown territory are highly probabilistic, and the success of predictions is usually low. *Nonetheless, there is scope for improvement in quality, process, and accuracy of geological predictions, which has also been dealt with in this study.*

(v) The high operating cost of deepwater rig necessitates minimization of nonproductive rig time. Nonproductive rig time is caused by various factors, which need to be identified and suitable measures be taken to minimize these. *Toward this goal, the current study deals with reduction in downhole complications, arresting delay in well completion and others.*

All these and more have been dealt with in this study.

4.2 OBJECTIVES

Due to high risk and exorbitant cost of deepwater exploration, E&P companies are under pressure for minimizing cost, maximizing utilization of available resources and opportunities for discovery of precious oil and gas.

The present exercise aims to improve the efficacy of some G&G activities and optimize G&G strategies taking into consideration techno-economic factors. The objectives of this study are as follows:

- Optimize G&G strategies for deepwater oil and gas exploration.
- Identify, assess, and improve the efficacy of some G&G activities and decisions, which are not as visible as other E&P activities and are not closely monitored.

- Analyze their effect on well economics through the following:
  (a) Optimize G&G evaluation time of deepwater wells especially wireline logging and identify areas for improvement.
  (b) Examine techno-economics of using LWD in own (deepwater) rig.
  (c) Improve the accuracy of geological predictions.
  (d) Effect on downhole complications of a well due to the variation in formation pressure (between predicted and actual).
  (e) Arrest slippage in well completion time of deepwater wells.
  (f) Influence of people's factors on the success and performance of deepwater exploration.

This exercise was carried out in an E&P company, which had embarked on deepwater exploration with lots of hopes, as its success could change the fortune of the company and bring prosperity to the country. The organization had created a special deepwater exploration project with dedicated resources, which were attached to a large offshore Asset of the company. The geographical extent of deepwater exploration undertaken by the organization covers a vast area extending to east and west coasts of the country, which are far apart and have water depth ranges between 1400 and 9900 ft.

The term "deepwater" generally means E&P activities beyond 400 m isobath or water depth in sea, ocean, or waterbodies. Earlier, even 200–300 m water depth was considered as deepwater. But with the advancement of technology, the ability to carry out E&P businesses at greater water depth has improved nowadays. Broadly speaking, water depth of 1320–5000 ft is termed deepwater (also called mid-water by some people), and water depth beyond 5000 ft is considered to be ultra-deepwater. However, there is no universal definition on this, and these ranges vary based on convenience of individual/groups.

4.3 METHODOLOGY

The study develops a framework for improving the aforementioned G&G activities and optimizing strategies for deepwater exploration. It takes into account G&G activities associated with formation evaluation such as wireline logging, coring, and well testing operations, which are required for completing a well and fulfilling the objective of assessing HC potential of prospects. The methodology followed in this study is further elaborated as follows:

(a) The data of 15 deepwater wells drilled till date were collected and analyzed. Since all these are exploratory wells, PT was not carried out in some of the wells based on the results of formation evaluation. Wireline logging was carried out in all these wells. Out of 15 deepwater wells, 6 are in west coast and 9 in east coast (Fig. 4.1).

Of these 15 deepwater wells, 7 wells were drilled by contract rigs, and the remaining 8 wells by own rigs (Fig. 4.2). All wells were drilled to their desired target depth with ±5% variations.

(b) The relevant data and information on G&G evaluation time including wireline logging, coring, and PT, well completion and rig cycle time, downhole complications, formation pressure at different depths, HC shows, and so on, for all deepwater wells drilled till date were collected and analyzed.

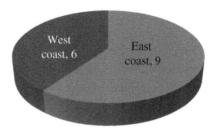

FIGURE 4.1 Number of wells studied: area-wise.

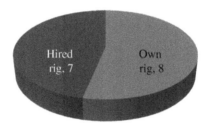

FIGURE 4.2 Number of wells studied: operatorship.

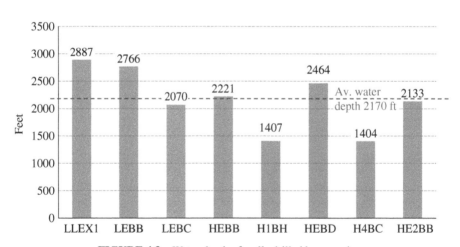

FIGURE 4.3 Water depth of wells drilled by own rigs.

(c) Own drillship is capable of drilling up to 3000 ft of water depth; beyond that hired rigs are deployed. The average water depth of wells drilled by own rigs is 2170 ft and ranges between 1400 and 2900 ft (Fig. 4.3).

Similarly, the average water depth of wells drilled by hired rigs is 6600 ft and ranges between 3500 and 9900 ft (Fig. 4.4).

In persuasion with the stated objectives of this study, the effect of some G&G activities and decisions on economics of deepwater wells is illustrated in the following paragraphs.

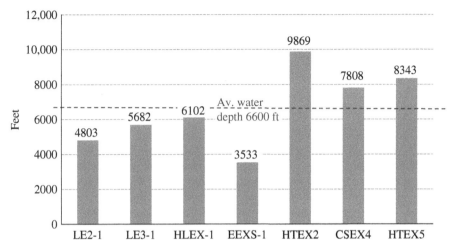

FIGURE 4.4   Water depth of wells drilled by hired rigs.

4.4   OPTIMIZATION OF G&G EVALUATION TIME IN DEEPWATER WELLS

G&G evaluation time is generally the time spent on wireline logging, coring, and PT. The evaluation time in this study is mostly due to wireline logging. Therefore, before we proceed further, a brief note on wireline logging, LWD, coring, and PT is presented in the following paragraphs, which will be useful to readers:

Conventional Wireline Logging: This is done to study the geophysical properties of the borehole/subsurface. It is basically a continuous recording of geophysical parameters in the borehole with respect to depth. Wireline logging is carried out to evaluate HC potentiality of the formation. It is done by lowering logging tools to the bottom of the hole or at the zone of interest and then pulling up (slowly) at the desired rate and recording geophysical attributes in high resolution with high accuracy. Logs that are normally taken with wireline logging for both open and cased holes are spontaneous potential (SP), gamma ray, resistivity, caliper, neutron, sonic, formation imaging, pressure and fluid sampling, wellbore seismic, magnetic resonance, and so on. The wireline tool is small and delicate compared to LWD and is powered through cable. It has high data transmission speed, susceptible to hole condition and problematic at high deviation wells.

LWD: This is reportedly capable of measuring formation properties and related parameters during drilling (real time), which are generally done by wireline logging except coring and sidewall coring. It records and transmits real-time data on formation properties that are of interest to drillers and G&G professionals. LWD helps in real-time decision making and controlling the well and is very useful to drill deviational and horizontal wells. It has the advantage of measuring formation properties before drilling fluid invades deeply into formation. The data speed is somewhat slow and is transmitted through mud pulse telemetry. LWD tool is large compared to wireline logging and is attached near the end of drill string as part of the bottom-hole assembly. Further discussion on LWD and wireline logging follows in Section 4.4.1.

Coring: This is done to study the properties of the drilled/reservoir rock, especially the porosity, permeability, angle of dip, rock characteristics, formation thickness, evidence of oil or gas, water, and so on. It gives many useful information and insights of the formation properties and lithology. A core sample is cylindrical in shape and is usually taken with the help of core bit, core barrel, and core catcher. There are different types/methods of coring, namely, (i) conventional coring; (ii) wireline coring, which is an offshoot of conventional coring; and (iii) sidewall coring.

Production Testing: After the well is drilled, it is tested for the presence of oil and gas. It is done by perforating the casing pipes at the desired depth (zone of interest) with the help of perforating guns. The formation fluid is allowed to flow in a controlled way. The drilling fluid (mud) is gradually replaced with water/brine, thus progressively reducing the hydrostatic pressure against the formation pressure. Formation fluid flows into the well, when the formation pressure exceeds the hydrostatic pressure of the well. The fluid flow is controlled through choke and bin. The main objective of PT is to evaluate reservoir flowing pressure, rate, water cut percentage, gas/oil ratio, gas/liquid ratio, well productivity, and other reservoir parameters.

4.4.1 Observations and Deep Dive

Seven deepwater wells were drilled by hired rigs in three different basins, namely, Basin 1 (two wells), Basin 2 (three wells), and Basin 3 (two wells) at a higher water depth ranging from 3500 to 9900 ft with the average being 6600 ft. The hired deepwater rig is equipped with LWD, which aids real-time measurements and continuous monitoring of various important parameters while drilling. The following key observations are made and their reasons have been explained:

(a) G&G evaluation (wireline logging and coring) time as percentage of well completion time is found to be higher in wells drilled by contract rigs compared to own rigs. There are wide variations ranging from 5.2 to 71.6% on account of logging and coring with the average evaluation time of approximately 31%. The evaluation time on account of wireline logging and coring in two out of seven wells drilled by hired rigs is very high, 71.6 and 64.3% of well completion time (Fig. 4.5; the *X*-axis shows well name and water depth).

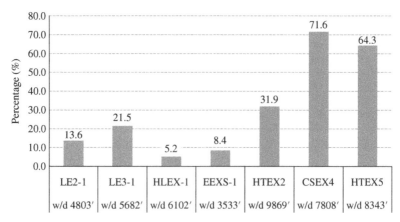

FIGURE 4.5 Evaluation (logging and coring) time as percentage of well completion time in hired rigs.

The reasons for such high percentage of evaluation time are less drilling time taken to reach the target depth and relatively higher evaluation time. In many wells, drilling days were found to be less than the evaluation (wireline logging and coring) days. It may also be noted that riserless drilling was resorted to few ultra-deepwater wells (rigs) with water depth around 9000 ft, which reduces the drilling time on account of not lowering the riser and related accessories. But it may lead to longer period of formation evaluation activities in the absence of cuttings and certain subsurface information for which multiple sidewall cores (SWCs) are taken.

Other factors responsible for high evaluation time are logging tool stuck-up, tool failures and repair jobs, string complications, hole enlargement and tripping for facilitating logging tool to pass through, insufficient hole cleaning and conditioning, and so on. Many of these nonproductive times (NPTs) may be avoided or substantially reduced by proper conditioning of well before lowering the logging tools and accessories.

It is also found that there is a lack of interface and coordination between drilling engineers and G&G professionals. Sometimes, there are varying opinions on different issues including NPTs. Therefore, there is a need to improve coordination and strengthen the interface between drilling engineers and G&G professionals.

(b) Total evaluation (logging, coring, and PT) time as percentage of well completion time in hired rigs is shown in Figure 4.6. No PT was carried out in six out of seven wells drilled by hired rigs. As a result, total evaluation times in these wells are limited to wireline logging and coring, which are same as that shown in Figure 4.5. PT was carried out only in one well (in Basin 1), which consumed 34.8% of well completion time (besides wireline logging and coring accounting for 13.6%). No PT was done in the remaining six wells drilled in Basin 2 and Basin 3, which appear to be less promising.

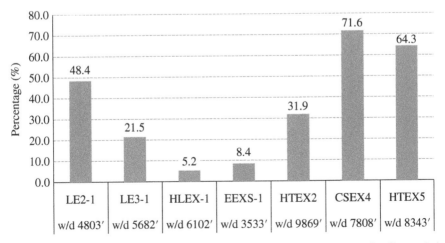

FIGURE 4.6 Total evaluation (logging, coring, and PT) time as percentage of well completion time in hired rigs.

The total evaluation time as percentage of well completion time in hired rigs is found at an average of approximately 36%. The reasons for high evaluation time on account of PT are similar to that of logging and coring, which has been deliberated in the previous paragraph. This has been further explained in Section 4.5.1b.

(c) **Total evaluation time** is the time spent on wireline logging, coring, and PT for various G&G measurements and studies in a well. Usually, evaluation time is higher in exploratory wells than development wells, as more time is spent on testing and collecting subsurface information in exploratory wells. **Well completion** time is the time between spud of a well and rig release. It is the sum of drilling time and evaluation time including wireline logging, coring, and PT in a well. **Cycle time** is the time between rig release from the earlier well and rig release of the current well. In other words, it is the sum of time taken for rig move, drilling, and evaluation including logging, coring, and PT. The breakup of rig cycle time of hired rigs is shown in Figure 4.7.

(d) Wireline logging for various G&G measurements was carried out in all wells drilled by hired rigs. Most of these parameters were also measured with LWDs, as G&G experts consider both LWD and wireline logging necessary to minimize risk. *The measurement of various parameters and tests carried out by wireline logging in all deepwater wells drilled till date apparently conforms to evaluations made by LWD.* The petrophysicists consider LWD as a real-time qualitative evaluation tool but don't entirely rely on evaluation based on LWD measurements, even though LWD is reportedly capable of replicating all major wireline logging measurements, such as gamma ray, density and photoelectric index, neutron porosity, borehole caliper, resistivity, sonic, borehole images, formation pressure, nuclear magnetic resonance, seismic while drilling, and so on, with the exception of coring and sidewall coring.

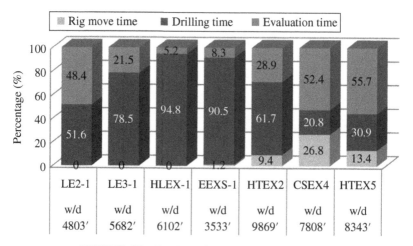

FIGURE 4.7   Breakup of rig cycle time: hired rigs.

However, it is claimed that the recent advancement in LWD suite takes care of this limitation to a great extent.

(e) Both LWD and wireline logging offer the same measurements *with some differences in quality, accuracy, resolution, and coverage.* Sometimes, some logs may not be directly comparable because of differences in the response characteristics of the measurements and the environment. However, corrections for instrument response variations and the environment are reportedly available for both LWD and wireline logging tools. LWD has advantages over wireline logging for providing real-time information and taking quick decisions based on various measurements such as porosity, resistivity, acoustic wave form, hole directions, geosteering, weight on bit, qualitative applications (picking formation tops), coring and casing points, and so on.

(f) But G&G experts are not in favor of replacing wireline logging with that of LWD and consider using both as necessary to minimize risk, especially in deepwater wells. The reasons generally cited by them are accuracy of measurements; besides that, the data transmission in LWD is through mud pulse telemetry while that of wireline logging is digital. Furthermore, the LWD logs are as per drillers' depth, while the depth accuracy of wireline logs is considered more reliable.

(g) In view of the earlier text, *this study considers that both LWD and wireline logging are necessary for deepwater wells* at the current level of technological maturity. *But there is scope for further trade-off and to optimize evaluation time on account of wireline logging.* The number of deepwater wells drilled so far is not sufficient to arrive at a conclusive result, but it gives a fairly good idea about the need and scope for optimization of evaluation time. With more and more deepwater wells drilled in the same field or in the vicinity, it is expected that G&G evaluation time (especially on account of wireline logging) will come down and stabilize. *Considering the exorbitant hiring cost of deepwater drilling rigs, it is important that G&G evaluation time is optimized and the possibility of extended period of G&G tests is minimized and unnecessary tests are dispensed with.*

(h) It is well known and widely understood that G&G decisions are based on scientific facts, data, and modeling work, but a certain degree of subjectivity is associated with the interpretation of subsurface data, model, and studies. Experience plays an important role for the success of exploration work, which cannot be replaced by shortcuts. However, a certain degree of accountability is to be instilled without curbing the creative freedom, as it is done in other disciplines.

In the instant case, G&G decisions for carrying out wireline logging for extended period in all deepwater wells drilled during 1 year raises eyebrow, especially when costly LWDs are deployed in these rigs and the findings of LWD are conformed by wireline logging in each and every case. Justifications for this extended period of evaluation also raise the question of whether sufficient due diligence was carried out for the selection of these deepwater prospects. G&G professionals need to introspect on this, in order to improve the efficacy of their decisions.

4.4.2 Suggestions

Based on the aforementioned findings and discussions in Section 4.4.1, the following suggestions are made to optimize evaluation time in deepwater wells:

(a) Evaluation based on LWD measurements may be relied upon beyond the quick look analysis, and further evaluation by wireline logging may be planned accordingly. There is scope for reduction or optimization of evaluation time on account of wireline logging and coring. If LWD measurements suggest nonpromising results, only then wireline logging may be decided judiciously. Evaluation time in a well may be optimized, and the extended period of evaluation on account of wireline logging especially when LWD measurements are nonpromising should justify its cost.

(b) While planning and designing a well, due importance may be given on techno-economic factors and issues. It should not be based on technical issues alone; multidisciplinary team (MDT) should discuss related economic issues and weigh their decisions on techno-economic perspectives.

(c) In order to minimize the chances of logging/PT tool stuck-up, string complications, tool failures, and related NPTs, *due care may be taken for proper conditioning of well before lowering logging/PT tools and accessories.* The old adage "drill a hole, clean a hole" may be followed that would help in reducing avoidable nonproductive drilling time.

It is also impressed upon to *improve coordination and strengthen the interface between drilling and well completion/G&G professionals to minimize NPTs* on account of these.

(d) To reduce the chances of failure, G&G data may also be evaluated by independent expert or third party for confirmation of major G&G decisions.

4.4.3 Benefits

If decisions were taken with due consideration of LWD measurements and *optimizing evaluation time on account of wireline logging*, it could have been possible to reduce the current evaluation time by at least 50%. In other words, the average wireline logging evaluation time that is approximately 31% at present can be reduced to 15.5% of well completion time. As more and more deepwater wells are drilled and experience gained on this count, it would not be difficult to achieve this. This means that around **53 rig-days** (i.e., 50% of 106 rig-days spent on evaluation) amounting to **USD 26.5 million** (i.e., 106 rig-days × USD 500,000 per day operating cost) could have been saved in a span of 1 year. In other words, additional **2–3 wells** (considering an average well cost of USD 9–14 million) could have been drilled for finding new oil and gas during this short period.

4.5 ECONOMICS OF ACQUIRING LWD FOR OWN DEEPWATER RIG

4.5.1 Observations and Deep Dive

Eight wells were drilled by own rig in Basin 1 (seven wells) and Basin 4 (one well) at a lesser water depth ranging between 1400 and 2900 ft. The deepwater drilling capacity of own rig is limited and is not equipped with LWD. The hiring cost of LWD is quite high and the operating cost of own deepwater rig is much less than that of comparable hired deepwater rig. The following important observations are made and their reasons are explained:

(a) The average G&G evaluation (wireline logging and coring) time as percentage of well completion time by own rig is found to be consistent at approximately 7% and is shown in Figure 4.8.

(b) Total evaluation (wireline logging, coring, and PT) time as percentage of well completion time is relatively high at 25% in wells drilled by own deepwater rig, but it's less than that of hired rigs (~36%). PT was carried out in all wells drilled by own deepwater rig (in Basin 1 and Basin 4), except one. PT alone consumed an average of 18% of well completion time.

The total evaluation time as percentage of well completion time of all wells drilled by own rig is shown in Figure 4.9. Basin 1 seems to be promising as PT was conducted in all wells (except one) and traces of HC shows were observed in some of these wells.

The evaluation time for PT primarily depends on number of objects and zones to be tested. The high evaluation time are also due to PT tool stuck-up, tool failures and repair jobs, string complications, detailed testing, insufficient hole cleaning and conditioning, and so on, which are similar to wireline logging and coring as discussed earlier. *NPTs on account of these may be considerably reduced by proper conditioning of well prior to lowering of PT tools and accessories.*

The breakup of rig cycle time of own rig is shown in Figure 4.10 (the *X*-axis shows well name and water depth).

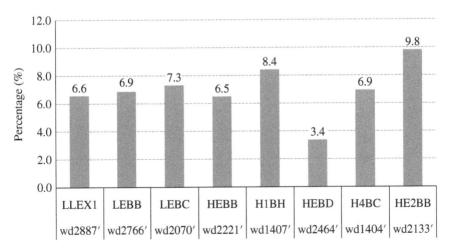

FIGURE 4.8 Evaluation (logging and coring) time as percentage of well completion time in own rig.

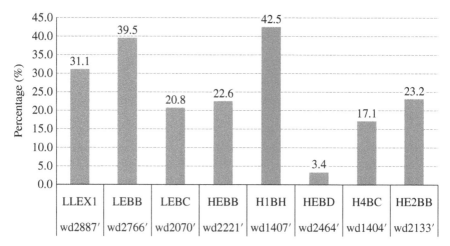

FIGURE 4.9   Total evaluation (logging, coring, and PT) time as percentage of well completion time in own rigs.

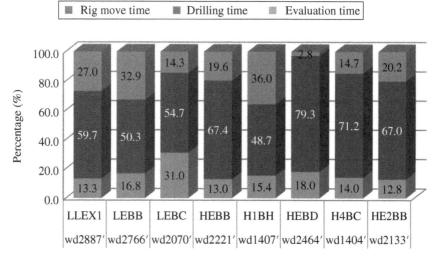

FIGURE 4.10   Breakup of rig cycle time: own rig.

4.5.2   Suggestions

Based on the findings and discussions in Section 4.5.1, the following suggestions are offered on the techno-economics of using LWD in own deepwater rigs:

(a) LWD may help in saving considerable amount of avoidable NPT, but its deployment should be governed by technical and economic considerations. The use of LWD will be competitive in own rig, if the current hiring rate of LWD is reduced by half. LWD in deepwater exploration can predict abnormal subsurface condition and provide valuable information in real time to avoid well complication, which may save substantial rig time on account of NPT. Therefore, *the use of LWD in own rig may be decided judiciously based on operational need and economic factors including trade-off between rig time saving potential and LWD hiring cost.*

Own deepwater drilling rig is capable of drilling up to 3000 ft water depth, which is comparatively low than that of contract rig (capacity) that are operating at ultra-deepwater. Considering low operating cost of own rigs, geological condition, and past experience of drilling in this area, it is felt that there are less chances of surprise. Therefore, the use of LWD at such low water depth (1400–3000 ft) for own rig is not economic at the current hiring rate. Instead, wireline logging and other G&G tests are found to be cheaper.

4.6 IMPROVE ACCURACY OF GEOLOGICAL PREDICTIONS

4.6.1 Observations and Deep Dive

The following key observations are made on geological predictions:

(a) Twenty-eight HC shows were predicted in six wells at west coast, against which **only one HC show** was observed (Fig. 4.11). The efficacy of prediction is not encouraging.

It raises the question of whether sufficient due diligence satisfying the necessary G&G criteria was carried out before releasing these prospects. It is believed that sometimes low-priority well locations are released to continue the drilling program and to avoid idling of costly deepwater rigs.

Furthermore, the present case of deepwater and ultra-deepwater exploration in this area is purely wildcat in nature with limited data, information, and support from the adjacent blocks/prospects. The predictions are mostly based on seismic data, which have limitations. These might be the reasons for low success of predictions in west coast.

(b) In east coast 105 no. of HC shows were predicted in nine wells, against which 62 no. of HC shows were observed in seven wells (Fig. 4.12). The success rate or efficacy of prediction seems to be promising.

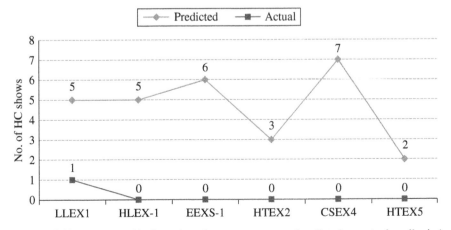

FIGURE 4.11 Number of hydrocarbon shows at west coast (predicted vs. actual: well-wise).

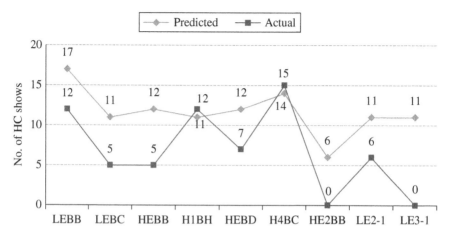

FIGURE 4.12 Number of hydrocarbon shows at east coast (predicted vs. actual: well-wise).

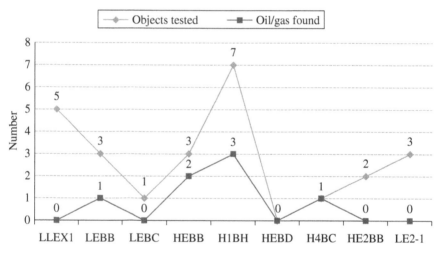

FIGURE 4.13 Number of objects tested versus oil/gas found (well-wise).

(c) PT was carried out at 25 objects in nine wells (east coast—8 and west coast—1), but traces of gas were found from seven objects in four wells (i.e., success ratio of 1 : 3.5). No commercial discovery has been made till the date of this study (Fig. 4.13).

4.6.2 Suggestions

Based on the findings and discussions in Section 4.6.1, the following suggestions would improve the accuracy of geological predictions:

(a) Rigorous due diligence fulfilling all necessary G&G criteria may be carried out for the selection of deepwater prospects and release of well locations for drilling. Many a times, low-priority wells having less geological chance factor (GCF) are released to avoid idling of rigs, especially deepwater rigs whose idling cost is exorbitantly high. This is not a healthy practice and requires deep introspection of the current practices followed, and invisible management policy (under the guise of operational need) is to be discouraged.

(b) The accuracy of geological predictions may be further improved through review by independent experts or third party and peer review. If this process is in place, then further qualitative improvement is impressed upon. It would not be out of place to mention that sometimes optimistic predictions are deliberately made to satisfy the techno-economic criteria for the release of well locations in order to avoid idling of costly rigs.

(c) The new and innovative technology-driven data acquisition and processing technique, use of real-time downhole tools, and knowledge gained through earlier exploration campaign in deepwater basin, and so on, are expected to provide more information and confidence to G&G professionals, which would improve the accuracy of geological predictions.

4.7 EFFECT ON DOWNHOLE COMPLICATIONS DUE TO THE VARIATION IN FORMATION PRESSURE

4.7.1 Observations and Deep Dive

The following important observations on downhole complications due to variations in formation pressure (between predicted and actual) are made and their reasons explored:

(a) The average downhole complications as percentage of drilling time is approximately 20% in wells drilled by own rigs and is approximately 18% in wells drilled by contract rigs.

Three out of eight wells drilled by own rig experienced downhole complications between 35 and 45% of drilling time despite operating in relatively low water depth of 2100–2500 ft compared to hired rigs (Fig. 4.14; the X-axis shows well name and water depth).

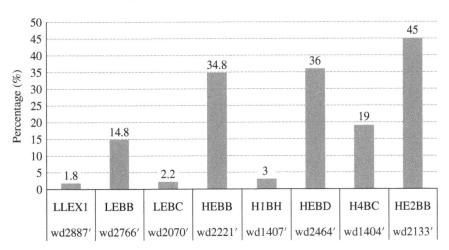

FIGURE 4.14 Downhole complications as percentage of drilling time at wells drilled by own rigs.

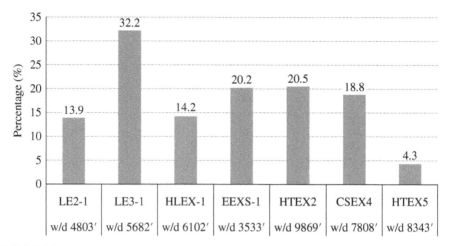

FIGURE 4.15 Downhole complications as percentage of drilling time at wells drilled by hired rigs.

The reasons for such high downhole complications encountered by own rigs are many, but prominent among them is supposedly "drilling practices." There is scope for improvement on this count (standard drilling practices) in own deepwater rig.

(b) The average downhole complications as percentage of drilling time is approximately 18% in wells drilled by contract deepwater rigs, which are operating in much higher water depth of 4800–9900 ft (with an average water depth of 6600 ft). One well (operating at a water depth of ~5700 ft) encountered downhole complications at 32% (Fig. 4.15). *The performance of contract rigs on this count is commendable as these rigs are operating at much higher water depth, and downhole complications usually increase with the rise in water depth.*

(c) Downhole complications at various pressure zones were correlated with the variation in formation pressure (i.e., pressure variation = predicted pressure − actual pressure).

The effects of pressure variation on downhole complications are shown in Figures 4.16 and 4.17 and explained as follows:

  (i) There is a positive correlation between formation pressure variation (predicted − actual) above 10% and downhole complications at the corresponding depth as indicated in Figures 4.16 and 4.17.

  (ii) Little or no downhole complications were found in zones where pressure variation is below 10%, suggesting pressure variation up to 10% may be acceptable.

  (iii) Even though there was no pressure variation in some pressure zones, downhole complications were observed at the corresponding depth implying that other factors are also responsible for downhole complication while drilling.

This may be further tested with additional well data to enhance the credibility of results and standards.

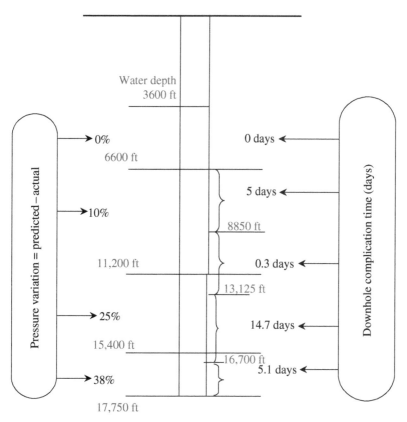

FIGURE 4.16   Well no.: EEXS-1.

4.7.2   Suggestions

Based on the findings and discussions in Section 4.7.1, the following suggestions would minimize downhole complications in deepwater wells:

(a) The downhole complications may be reduced through better synergy and information sharing among various groups and emphasizing the need of dissemination of useful technical information. In-depth analysis of available data may be undertaken to study the effect of formation pressure variations on downhole complications, so that the variation in predicted formation pressure is within the acceptable range.

(b) Furthermore, *the downhole complications may be reduced by following standard drilling practices. The unsafe drilling practices such as inadequate hole cleaning, very high rate of penetration (ROP) to meet drilling target, ignoring subsurface indications, and so on, need to be discouraged. Avoiding such unhealthy practices would improve drilling practices (quality) and minimize downhole complications paving way for efficient and productive well completion.*

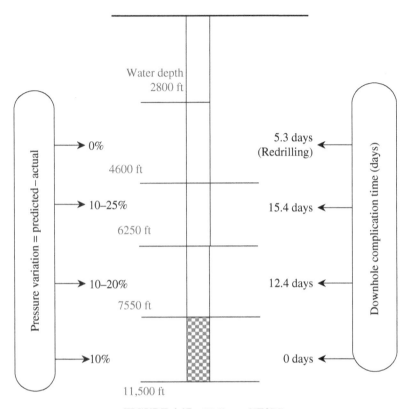

FIGURE 4.17   Well no.: HE2BB.

4.8   SLIPPAGE IN WELL COMPLETION TIME OF DEEPWATER WELLS

The following key observations are made on delay in deepwater well completion and their reasons.

4.8.1   Observations and Deep Dive

(a) The average well completion time by own rigs is 31.5% higher than its planned completion time in deepwater wells. In other words, an average 31.5% slippage is observed over planned well completion time in own rigs. However, one well was completed 11% ahead of scheduled completion, and another three wells were completed with marginal slippage. But half of the wells experienced high slippage ranging between 36.1 and 104.5% (Fig. 4.18; the X-axis shows well name and water depth).

The delay was mainly due to downhole complications, prolonged PT and related complications, and other factors such as work practice, work culture, and so on.

(b) More than half of deepwater wells were completed by hired rigs before the planned completion time. Three wells were completed marginally ahead of schedule, and one well was completed approximately 43.5% earlier than the planned completion time. These are due to faster drilling, the decision of not

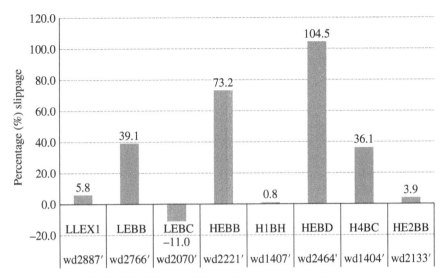

FIGURE 4.18 Slippage (%) in well completion time by own rigs.

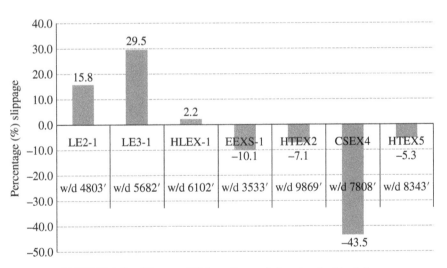

FIGURE 4.19 Slippage (%) in well completion time by hired rigs.

carrying out certain planned jobs such as testing, lowering of casing upon reaching the target depth, and so on.

Nonetheless, *this may be a lesson for own rig crews as hired rigs operated at much greater water depth compared to own rig*. The completion of one well was delayed by 29.5% for respud of well due to equipment failure. All wells were drilled to the desired target depth. The average slippage in well completion by hired rigs is found to be **negative**, that is, completed ahead of schedule. All these are depicted in Figure 4.19.

4.8.2 Suggestions

Based on the aforementioned findings in Section 4.8.1 and reasons thereof, the following suggestions would help in completing deepwater wells in time by own rigs:

(a) MDT must actively pursue, monitor, and ensure that no rig waits for want of tools, materials, and equipment.

(b) Learning from hired rigs may be shared among the crews of own rigs. Same failure at different locations may be avoided by sharing information among various groups.

(c) The cost–benefit analysis for buying or hiring of "nontypical" tools at short notice through third party may be worked out, considering the time frame of deepwater drilling program. For example, nontypical tools and accessories like high resolution or high pressure and high temperature (HPHT) are not readily available in the market and usually have long procurement lead time. So special arrangement for getting it quickly would be of immense help.

4.8.3 Benefits

The aforementioned measures would help in containing slippage in well completion by own rigs. Like hired rigs, if the actual well completion time of own rig is brought to the level of planned well completion time following the suggestions in this study, it would save around **228 rig-days** or **USD 22.8 million per rig-year** (i.e., 228 rig-days × at USD 100,000 per day operating cost).

4.9 INFLUENCE OF PEOPLE'S FACTORS ON THE SUCCESS OF DEEPWATER EXPLORATION

4.9.1 Observation

The current deepwater exploration setup lacks effective team integration akin to MDT of Assets. It does not have dedicated personnel from G&G, fluids, production, logistics, and others and lacks cohesion among experts from various groups. There is much scope for improvement in these areas, which may be pursued further.

4.9.2 Suggestions

The key to success of deepwater exploration lies beyond technology to "people's" factors of team integration, shared culture, effective processes, and work environment. These can be improved upon through the following measure:

1. The formation of MDT comprises dedicated representatives from G&G, drilling, cementing, logging, fluids, production, materials, logistics, and engineering, who would

    (i) Work in close proximity (and not in different offices or locations, as is the current practice) and discuss regularly among themselves;

(ii) Understand and appreciate each other's problems and issues, share information for performance improvement and building trust;

(iii) Take decisions based on wide consultations with experts from various groups and collective wisdom of the team.

4.10 BENEFITS

The measures suggested in this study (refer to Sections 4.4.2, 4.5.2, 4.6.2, 4.7.2, 4.8.2, and 4.9.2) covering economic, noneconomic, and operational areas would help in optimizing G&G strategies for exploring oil and gas in deepwater. This would result in considerable savings that may be plowed back to enhance further opportunity for discovery of oil and gas:

(a) *If decisions were taken with due consideration of evaluation based on LWD, which subsequently proved true for all deepwater wells, and optimizing evaluation time on account of wireline logging, it could save around 53 rig-days or USD 26.5 million per year at the current level of activity. This means two to three additional wells could have been drilled within this period, which would have increased chances of finding new oil and gas.*

(b) If own deepwater rig adheres to well completion time and contains the current slippage following suggestions offered in this study, it would save around 228 rig-days or USD 22.8 million per annum.

4.11 LIMITATIONS OF THE STUDY AND SCOPE FOR IMPROVEMENT

The credibility of this study would be further improved, if more well data are considered for assessing the efficacy of G&G activities and optimizing G&G strategies. The data for deepwater wells available at the time of this study were limited. As more and more deepwater wells have been drilled in subsequent days and years, further work may be done with additional well data, which would provide more realistic results.

Furthermore, with the availability of sufficient well data, deepwater wells may be further categorized into a number of categories, for example, deep wells (water depth 1320–5000 ft) and ultra-deep wells (water depths > 5000 ft). However, for this study, it is suggested to divide deepwater wells into three categories, namely, deep wells (water depth 1400–3500 ft), ultra-deep wells (water depth 3501–7500 ft), and super ultra-deep wells (water depth > 7500 ft).

Accordingly, G&G strategies for each category of wells may be optimized instead of a single deepwater category. Thus, optimizing G&G evaluation time, examining techno-economics of LWD deployment in own rig, effect on downhole complications due to variations in formation pressure, arresting slippage in well completion, and so on, may be undertaken for each category of deepwater wells (i.e., deep, ultra-deep, and super ultra-deep). This would provide more specific and realistic results, which would help in improving the performance of G&G activities and optimizing G&G strategies for each category of deepwater wells.

4.12   CONCLUSION

Deepwater exploration is a promising frontier for future oil and gas reserves, but it is costly and challenging. E&P companies across the world are hard pressed to contain cost, maximize available resources, and find opportunities for deepwater exploration. They are continuously monitoring the performance of various E&P activities, but some G&G activities and decisions escape attention due to its inherent subjectivity and uncertainties. The present study aims to identify, assess, and improve the efficacy of these G&G activities and strategies based on techno-economic factors. Some key observations and **suggestions** for optimizing G&G strategies in deepwater exploration are summarized in the following paragraphs:

(i) G&G evaluation time in deepwater wells is found to be very high. There is scope for optimization of evaluation time on account of wireline logging, especially when LWD is used. LWD results of all wells in this study have been confirmed by wireline logging; therefore, due importance may be given on LWD results. Further measurements using wireline logging may be decided judiciously. Nevertheless, G&G decisions may be taken on techno-economic considerations, instead of technical factors alone, which would help in optimizing G&G strategies for deepwater exploration.

(ii) Ensure proper hole cleaning before lowering of logging/PT tools and accessories, which would minimize the chances of tool stuck-up and failures, string complications, and so on, which are responsible for extended period of evaluation activities. Also, improve coordination and strengthen the interface between drilling and well completion/G&G professionals to minimize wireline logging and PT-related NPTs.

(iii) To reduce the chances of failure, G&G data may also be evaluated by independent expert or third party for confirmation of major G&G decisions.

(iv) The use of LWD in own rig should be governed by operational need and economic consideration. Use of LWD in own rig is not economic at the current hiring rate considering geological condition and past experience in the areas of operation, low operating cost, and capacity of own rig, which is operating at relatively lesser water depth. Instead, wireline logging is found cheaper and preferred.

(v) Proper due diligence may be conducted for release of (deepwater) well locations. The practice of optimistic prediction may be replaced with most likely prediction focusing on accuracy and geological chance factors rather than concern for idling of deepwater rig.

(vi) The accuracy of geological predictions may be improved through review by independent experts or third party and peer review. Further, improvement is possible with the induction of innovative technology for data acquisition and processing, using real-time downhole tools and experience gained through earlier exploration campaign in deepwater basins.

(vii) The downhole complications may be reduced by following standard drilling practices, better synergy, and information sharing among various groups and analyzing the effect of formation pressure variations on downhole complications.

(viii) The well completion time of deepwater wells drilled by own rig may be reduced through better monitoring, learning from failures and from contract rigs, sharing lessons, improving work practices and work culture, and so forth.

(ix) Finally, the success of deepwater exploration lies beyond technology to "people's factors" of team integration, shared culture, effective processes, and work environment, which can be improved through the measures suggested in this study.

(The detailed suggestions are presented in Sections 4.4.2, 4.5.2, 4.6.2, 4.7.2, 4.8.2, and 4.9.2.)

The study develops a framework for optimizing G&G strategies in deepwater exploration, which can be continuously improved and updated with newly acquired well data. The efficacy of G&G activities/decisions would improve and the objective of optimizing G&G strategies for deepwater exploration would be met, if these suggestions are considered. These recommendations have potential of saving around USD 49.3 million in less than a year.

The importance of offshore supply vessel (OSV) for smooth and uninterrupted offshore operations is well known, which has been mentioned in earlier chapters. Chapter 5 develops a waiting line model to determine the optimum number of OSVs required for supply of materials to various offshore rigs, platforms, and installations, so that the waiting time for materials is minimized and the system cost is optimized.

REVIEW EXERCISES

4.1 What are the various G&G evaluations that are normally carried out in a deep(water) well? What are the purposes of these evaluations?

4.2 How would you optimize G&G evaluation time in deepwater wells?

4.3 As a geologist, how would you plan to take a core sample? What are the different types of coring? State their purpose and limitation.

4.4 What are the basic differences between wireline logging and LWD?

4.5 What are the commonly used logs that are generally referred to by E&P professionals? State their purpose and utility.

4.6 As a G&G specialist, how would you improve accuracy of geological predictions?

4.7 What are the reasons for slippage in the completion of deepwater well as per this study? As a drilling specialist, how would you plan to minimize these delays?

4.8 What do you understand by people's factors influencing deepwater exploration? How important is it for the success of deepwater exploration?

Advance

4.9 What are the major causes of downhole complications in deepwater wells? As a drilling expert, what would you recommend to minimize it?

4.10 What do you understand by shale problem? What are the reasons and remedial measures?

4.11 What are the different types of well completion? State their advantages and limitations. As in charge of well completion team, how would you plan well completion?

FURTHER READING

Dewan, J.T., Essentials of Modern Open-Hole Interpretation, PennWell Books, Tulsa, OK, 1983.

Ellis, D.V. and Singer, J.M., Well Logging for Earth Scientists, Springer, Dordrecht, 2007.

Gatlin, C., Petroleum Engineering, Drilling and Well Completion, Prentice-Hall Inc., Englewood Cliffs, NJ, 1960.

Home, R.N., Modern Well Test Analysis, 2nd edition, Petroway Inc., Palo Alto, CA, 1995.

Lee, J., Well Testing, SPE Text Book Series, Vol. 1, Society of Petroleum Engineers of AIME, New York, 1982.

Leffler, W.L., Pattarozzi, R., and Sterling, G., Deepwater Petroleum Exploration & Production: A Nontechnical Guide, 2nd edition, PennWell Publishing Company, Tulsa, OK, 2011.

Prison, S.J., Handbook of Well Log Analysis, Prentice-Hall Inc., Englewood Cliffs, NJ, 1963.

Rider, M.H. and Kennedy, M., The Geological Interpretation of Well Logs, 3rd edition, Rider-French Consulting Limited, Sutherland, 2011.

Serra, O., Fundamentals of Well-Log Interpretation: The Interpretation of Logging Data, Developments in Petroleum Science, 15b, Elsevier Science Publishers, Amsterdam, 1987.

Tiab, D. and Donaldson, E.C., Petro-physics: Theory and Practice of Measuring Reservoir Rock and Fluid Transport Properties, 3rd edition, Elsevier, Boston, MA, 2012.

Toby, D., Well Logging and Formation Evaluation, Elsevier, Boston, MA, 2005.

Weimer, P. and Slatt, R.M., Petroleum Systems of Deepwater Settings, Distinguished Instructor Series, Society of Exploration Geophysicists (SEG) and European Association of Geoscientists and Engineers (EAGE), Tulsa, OK, 2004.

Wylie, M.R.J., The Fundamentals of Well Log Interpretation, 3rd edition, Academic Press, New York, 1983.

Useful Links

Bastos, A.R.G., Logging While Drilling versus Conventional Wire Line: Comparison, available at: https://fenix.tecnico.ulisboa.pt/downloadFile/395145922699/TRADU%C3%87%C3%83O1.pdf (accessed on February 29, 2016).

High-Resolution LWD Image Logs versus Wireline Image Logs, Offshore Magazine, available at: http://www.offshore-mag.com/articles/print/volume-67/issue-5/drilling-completion/high-resolution-lwd-image-logs-versus-wireline-image-logs.html (accessed on February 29, 2016).

Lamont-Doherty Earth Observatory (The Earth Institute of Columbia University), February 14, 2008, An Introduction to Logging While Drilling, Seminar in Marine Geophysics, available at: http://www.ldeo.columbia.edu/res/div/mgg/lodos/Education/Logging/slides/LWD_Feb_15_2008.pdf (accessed on March 18, 2016).

Page, S., Can LWD Replace Wireline?, available at: http://www.wellservicingmagazine.com/cover-story/2006/09/can-lwd-replace-wireline/null (accessed on February 29, 2016).

5

OPTIMIZATION OF OFFSHORE SUPPLY VESSEL FLEET SIZE USING QUEUING THEORY

5.1 INTRODUCTION AND BACKGROUND OF THE PROBLEM

The importance of offshore supply vessel (OSV) for supply of materials to offshore installations and the necessity of having sufficient number of OSVs for carrying out uninterrupted exploration and production (E&P) activities are well known, and the impact of waiting for OSVs has been discussed in Chapters 2 and 3. The current study formulates an optimization model to determine optimum OSV fleet size, so that the waiting time for supply of materials to various offshore installations is minimized.

An offshore oil field is characterized by numerous installations such as production platforms, drilling/workover rigs, and others, which are operating round the clock for realization of its stated objectives, targets, and tasks. Offshore rigs are of different types such as jack-up rig, semisubmersible rig, and drillship, which are primarily engaged for drilling of wells, but are also used for workover jobs. Based on the well types, drilling is categorized as "exploratory" and "development." Exploratory drilling is aimed at finding new oil and gas, whereas development drilling is associated with enhancing production capacity from the existing field. Oil produced at various wells is gathered at the production platforms where oil and associated gas are separated, before pumping to onshore facilities through pipelines for further processing.

Offshore rigs and production platforms are self-contained units where sets of crews are stationed round the clock. Usually, the crew members work on 28 days on/28 days off, or 56 days on/28 days off, or any other pattern depending on the need, convenience, economics, and rules of the company.

In order to ensure smooth E&P activities, it is necessary to supply materials, such as fuel, cement, mud chemicals, drilling and casing pipes, tubular, equipment, spare parts, tools,

Optimization and Business Improvement Studies in Upstream Oil and Gas Industry, First Edition.
Sanjib Chowdhury.

water for drilling and drinking, and other items to offshore installations. These materials are supplied by OSVs, which are specially designed cargo ship built exclusively for this purpose. OSVs are lifeline of offshore operations; therefore, it is essential that sufficient numbers of OSVs are available for supply of materials to offshore installations. An OSV round trip consists of loading of materials, dispatch delay, transportation, and unloading of materials. During a round trip, it generally supplies material to a number of destinations/installations, and a round trip time depends on the distance of these installations from the base. OSVs can be dispatched at any point of time during day and generally do not sail off from the port at night.

Offshore E&P activities are costly and risky. The waiting cost of drilling rigs and production installations is exorbitantly high; therefore, installations waiting for OSVs is undesirable. This exercise uses queuing theory to estimate optimum OSV fleet size with the objective of optimizing waiting time of installations.

5.2 OBJECTIVES

The policy and priority of the management is that at no point of time, offshore installations shall wait for OSVs, as idling cost of drilling rigs and production platforms is extremely high. This implies that there shall be adequate number of OSVs to cater the need of offshore installations. But OSVs are costly capital item, whose operating cost is about one-fifth of offshore rig operating cost. Therefore, efficient utilization of OSVs is of prime importance. It is necessary to trade off between waiting time of installation and service cost of OSVs.

The objectives of the current exercise are as follows:

(a) To estimate optimum OSV fleet size, so as to optimize the waiting time of installations for materials
(b) To determine various performance measures and operating characteristics of the queue, such as waiting time in the queue and that in the system, queue length and system length, servers' utilization, probability of call waiting, and so on, under varying conditions to aid decision making

This study was carried out in an oil field, which is situated around 120–160 nautical miles away from the seashore, and rigs/installations are spread over an area of more than 90,000 km².

5.2.1 Methodology

This study uses queuing theory to determine optimum OSV fleet size for uninterrupted supply of materials to offshore installations. The methodology adopted for this exercise is as follows:

(a) The logistics for supply of materials to offshore installations by OSVs were studied, and the relevant data required for developing waiting line model were collected, which are discussed in Section 5.4.1.

(b) A queuing model was developed and the process of formulating the model is discussed in Section 5.4.

(c) The model was tested and analyzed with varying parameters such as number of OSVs and generating various operating characteristics for decision making, which are discussed in Section 5.5.

5.3   WAITING LINE MODEL: A BRIEF NOTE

Waiting line phenomenon is common in real life and is experienced by all. There is hardly anyone who has not experienced waiting in queue for bill payments, bank teller, and ATM; waiting for their turn in cafeteria, restaurants, and supermarket; waiting in hospital or health center for doctor; waiting in saloon for barber; waiting for bus, train, or other transport; vehicles waiting in gas station and tollbooth; trucks and ships waiting for loading and unloading; waiting at the immigration counter, and security check at the airport; aircraft waiting for takeoff; waiting for judicial process; and so forth. Queuing theory has extensive use in real world, and its application can be found across commercial, noncommercial, transport, service, health care, education, industrial, and other sectors. It is particularly useful for designing processes, such as traffic congestion in telecommunication, networks, and transportation (road, rail, air, and water); in manufacturing units (machine breakdown and repair, and scheduling of production control and inventory control); and so on.

Queuing theory is essentially mathematical analysis of waiting line problems in stochastic system. It is a quantitative technique to study the characteristics of queue and assess system performance. Each queuing problem is unique with distinct characteristics and is different from the other. However, most of them can be approximated and grouped under different categories based on well-defined mathematical criteria. Thus most of the waiting line problems can be expressed in mathematical form and solved with sound mathematical logic. It requires sound knowledge of mathematics and analytical skill.

Agner Krarup Erlang, a Danish mathematician and engineer, developed queuing theory in the early twentieth century to solve the telephone exchange/call congestion problem. Erlang's work is the basis of methodological framework of queuing theory using probabilistic methods, which also engineered the development of the present-day stochastic processes.

5.3.1   Components of Basic Queuing Process

Queues are formed when the short-term demand for service exceeds the capacity. The components of basic queuing process are as follows [1]:

Calling population: It is the source from which customers originate. The size of population may be finite or infinite, but the most commonly used population is infinite in nature. Population may be homogeneous (similar type of customers) or heterogeneous (different or mixed type of customers).

Arrival process: The input process of a queuing model is called **arrival process**. Arrival of customers is usually characterized by a probability distribution function, and the time between arrivals of two consecutive customers is called **interarrival time**. If more than one customer arrives at a given point of time, it is called **bulk** arrivals. If a customer decides not to join a queue (considering its length), it is called **balking**. If a customer leaves after waiting in a queue for some time, it is called **reneging**. If a customer switches queue, it is called **jockeying**.

Queue configuration: It indicates the number of queues in the system, that is, single or multiple lines for service. Different configurations are suitable for different purposes.

Queue discipline: It denotes the rule for service, that is, the order in which the customers in the queue will be served. The following rules are generally followed for serving: first-in–first-out/first-come–first-served (FIFO/FCFS), last-in–first-out/last-come–first-served (LIFO/LCFS), service in random order (SIRO), priority schemes (PS), and so on.

Service mechanism: The output process of a queuing system is called "service process". It is dependent on service facilities, that is, number and configuration of service channels (single or multiple servers), and characteristics of service time. The characteristics of service time are determined by studying and analyzing service time and process. Service time distribution may be general, exponential, hyperexponential, deterministic, and so on. Most of the queuing models are found having exponentially distributed service time.

5.3.2 Kendall's Notation

There can be various waiting line problems representing different conditions and processes. As a result, there are various types of queuing models, each representing a particular type of problems. Different authors and practitioners have used different notations for describing the queuing models. But the most popular among them is the Kendall's notation [2], which is denoted as **A/B/c/K/m/Z**, where

- A is the interarrival time distribution
- B is the service time distribution
- c is the number of servers
- K is the system's queue capacity
- m is the number of customers in the system
- Z is the queuing discipline

If only first three entries ($A/B/c$) are mentioned and the rest are omitted, then it is assumed that the source and queue capacity are both infinite, and the queue discipline is FIFO.
A and B may be denoted as

- GI for general independent interarrival time
- G for general service time
- E_k for Erlang-k distribution
- M for Markovian (memory less) or exponential distribution
- D for deterministic distribution
- H_k for hyperexponential (with k stages) distribution

There can be various combinations of finite-and infinite-source queuing systems. The most commonly used waiting line problems follow exponential (Poisson) distribution for interarrival time, exponential distribution for service time, multiple servers, infinite queue capacity, and FIFO queuing discipline. Various queuing models are denoted as follows:

- *M/M/c* queuing system: Markovian (Poisson) interarrival time and Markovian (exponential) service time with multiple servers.
- *M/M/*1 queuing system: is a special case of *M/M/c* model, where $c = 1$.
- *M/G/*1*/GD/∞/∞* queuing system: exponential interarrival time, general service time with single server, general discipline, infinite queue, and infinite system capacity.
- Some other combinations are *M/M/*1*/K*, *M/M/∞*, *M/M/c/∞/N*, *M/M/n/n*, *M/M/c/K*, *M/G/c*, *D/D/c/K/K*, GI*/M/c*, and so on.

5.3.3 Transient and Steady-State Condition

The queuing system essentially has two states of condition, namely, (i) **transient condition** and (ii) **steady-state condition**. If system behavior changes with time, it is said to be in transient condition. Such condition is generally common at the initial phase of operations of a new queuing system. The probability distribution of the transient state varies with time. On the other hand, the queuing system tends to settle down with the passage of time and reach to a steady sate of condition. The probability distribution of steady-state condition is stationary over time. It is easier to analyze and solve problems in steady-state condition compared to transient condition that changes over time.

5.3.4 Birth–Death Processes

The foundation of many queuing models is described in terms of birth–death process, wherein the **arrival of a customer** is referred to as **birth** and the **departure of a *served* customer** is called **death**. The flow line diagram of the birth–death process is shown in Figure 5.1, which explains the intricacies of queuing process.

In Fig. 5.1, node 1, 2, ... *n*, represents "state", that is, number of customers *n* in the system at a specified time *t*.

λ = arrival rate per unit time
μ = service rate per unit time
P_n = probability of *n* customers in the system
L_q = length of queue or number of customers in the queue
L = system length or number of customers in the system
W_q = waiting time in the queue
W = waiting time in the system

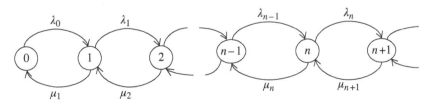

FIGURE 5.1   Flow line diagram of birth–death process.

5.3.5   Steady-State Analysis of Birth–Death Processes

The following conditions must be satisfied for the steady-state solution of birth–death processes [1]:

(a) The **mean arrival rate** of customers is **equal to the mean departure rate**

$$\text{i.e., Rate in}\left(\text{birth rate}\right)=\text{rate out}\left(\text{death rate}\right)$$

(b) The probability of being in one of the states must equal 1 (the sum of all probabilities is 1)

$$\text{i.e.,} \sum_{i=0}^{\infty} P_i = 1$$

(c) Meet the condition for existence of steady-state solution: $\dfrac{\lambda}{\mu} < 1$

The **balance equation** for each state is derived as follows:

State 0: $\mu_1 P_1 = \lambda_0 P_0 \rightarrow P_1 = \dfrac{\lambda_0}{\mu_1} P_0$

State 1: $\lambda_0 P_0 + \mu_2 P_2 = \lambda_1 P_1 + \mu_1 P_1 \rightarrow P_2 = \dfrac{\lambda_1}{\mu_2} P_1$

State n: $\lambda_{n-1} P_{n-1} + \mu_{n+1} P_{n+1} = \left(\lambda_n + \mu_n\right) P_{n-1} \rightarrow P_n = \dfrac{\lambda_{n-1}}{\mu_n} P_{n-1}$

Normalizing, we get $\sum_{i=0}^{\infty} P_i = P_0 \left\{ 1 + \dfrac{\lambda_0}{\mu_1} + \dfrac{\lambda_0 \lambda_1}{\mu_1 \mu_2} + \dfrac{\lambda_0 \lambda_1 \lambda_2}{\mu_1 \mu_2 \mu_3} + \ldots \right\} = 1$

Considering, $C_0 = 1$, $C_1 = \dfrac{\lambda_0}{\mu_1}$, $C_2 = \dfrac{\lambda_0 \lambda_1}{\mu_1 \mu_2} \ldots C_n = \dfrac{\lambda_0 \lambda_1 \ldots \lambda_{n-1}}{\mu_1 \mu_2 \ldots \mu_n}$

Steady-state probabilities:

$$P_0 = \frac{1}{\sum_{i=0}^{\infty} C_i} \quad \text{and} \quad P_n = C_n P_0 \tag{5.1}$$

The expected number of jobs in the system and in the queue (considering c parallel servers) is as follows:

$$L = \sum_{i=0}^{\infty} i \cdot P_i \quad \text{and} \quad L_q = \sum_{i=c}^{\infty} (i-c) \cdot P_i \tag{5.2}$$

The operating characteristics and system behavior of various types of queuing models may be derived following the birth–death process analysis in steady-state condition as described earlier. To know more about queuing theory and its application, interested readers may refer to books listed in "References/Further Reading."

5.4 FORMULATION OF THE MODEL

As per the stated objectives of this study, the current study develops a waiting line model for estimating optimum OSV fleet size for timely supply of materials to offshore installations, so that waiting time at various installations is minimized:

- The call for requirement of materials received at the base from offshore installations may be considered as *call arrival rate* (λ). These calls are received apparently at random as explained in Section 5.4.1 and assumed to follow the Poisson distribution.
- Based on the call, materials are dispatched to various destinations by OSVs usually on the FCFS basis. Therefore, OSVs may be considered as *server* and *the queue discipline* may be considered as FIFO/FCFS. There are a number of OSVs, and therefore, it is consistent with *multiserver channel*.
- An OSV usually touches one or more destinations before returning to the base. A round trip time of OSV may be considered as *service time* and the *service rate* (μ) is assumed to be exponential.

The aforementioned problem or scenario can be approximated to "*M/M/c*" category queuing model and can be solved as follows:

If c=number of servers (i.e., OSVs—working independently and performing similar operations)

n=number of calls in the system awaiting service or being served

Assuming, interarrival density: $\lambda e^{-\lambda t}$

and service time density for each server: $\mu e^{-\mu t}$

where individual service times are mutually independent.

Following the iterative process described in Section 5.3.5 (steady-state analysis of birth–death process) and solving the difference equations, the associated operating characteristics of a multiserver queuing model can be obtained as follows:

The probability that all servers are idle, that is, there is zero unit in the system (P_0) is as follows:

$$P_0 = \left[\sum_{n=0}^{c-1} \frac{1}{n!} \left(\frac{\lambda}{\mu} \right)^n + \frac{1}{c!} \left(\frac{\lambda}{\mu} \right)^c \frac{c\mu}{c\mu - \lambda} \right]^{-1} \tag{5.3}$$

The probability that a call waits is as follows:

$$\left(P_w \right) : P_w = \frac{\left(\lambda/\mu \right)^c}{c!\left(1 - \left(\lambda/c\mu \right) \right)} P_0 \tag{5.4}$$

The probability of n calls in the system (P_n) is as follows:

$$P_n = \begin{cases} \dfrac{1}{n!} \left(\dfrac{\lambda}{\mu} \right)^n P_0 & \text{for} \quad 1 \le n < c \\[3mm] \dfrac{1}{c! c^{n-c}} \left(\dfrac{\lambda}{\mu} \right)^n P_0 & \text{for} \quad n \ge c \end{cases} \tag{5.5}$$

The condition for the existence of steady-state solution is $\rho = \lambda/(c\mu) < 1$.
Queue length or average number of entities in the queue (L_q) is as follows:

$$L_q = \sum_{n=c}^{\infty} (n-c) P_n = \dots = \frac{(\lambda/\mu)^c \rho}{c!(1-\rho)^2} P_0$$

Solving,

$$L_q = \frac{\lambda\mu(\lambda/\mu)^c}{(c-1)!(c\mu-\lambda)^2} P_0 \tag{5.6}$$

As per Little's formula (which is a simple, useful, and general equation that does not depend on the arrival process distribution nor does it depend on the service process distribution and is also independent of number of servers and buffers in the system),

$$W_q = \frac{L_q}{\lambda}$$

and

$$L = \lambda W = \lambda \left(W_q + \frac{1}{\mu} \right) = L_q + \frac{\lambda}{\mu}$$

Applying Little's formula, the system length or average number of entities in the system (L) is as follows:

$$L = \frac{\lambda\mu(\lambda/\mu)^c}{(c-1)!(c\mu-\lambda)^2} P_0 + \frac{\lambda}{\mu} \tag{5.7}$$

The waiting time in the queue (W_q) is as follows:

$$W_q = \frac{\mu(\lambda/\mu)^c}{(c-1)!(c\mu-\lambda)^2} P_0 \tag{5.8}$$

The waiting time in the system (W) is as follows:

$$W = \frac{\mu(\lambda/\mu)^c}{(c-1)!(c\mu-\lambda)^2} P_0 + \frac{1}{\mu} \tag{5.9}$$

Server utilization:

$$\rho = \frac{\lambda}{c\mu} < 1 \left(\text{which is also known as system load or traffic intensity} \right) \tag{5.10}$$

5.4.1 Model Calibration

The data collection and model calibration is an important step for the success of waiting line model. Prior to data collection, the entire OSV operating system starting from receiving of calls, loading of materials, dispatch to destinations, journey period, and unloading of materials were studied in detail. The call logs, dispatch register, and round trip details were critically examined. The data were collected for a period of 12 months.

The requests for supply of materials at installations are consolidated and the dispatch schedule is prepared daily based on (i) whether OSV capacity is fully utilized or maximized during each trip, (ii) clubbing of materials for installations that are in the close vicinity or zone, and (iii) urgency of materials. *Therefore, the number of OSV sailing-off per day may be considered as effective call arrival rate (λ).*

- Number of trips made in a year $= 1161$
- Total round trip time in a year $= 9300$ days

- Number of calls received per day $= \dfrac{\text{number of trips made in a year}}{365 \text{ days}}$

- That is, average call arrival rate $\lambda = \dfrac{1161}{365} = 3.18$ per day

- Average service time $= \dfrac{\text{total round trip time in a year}}{\text{number of trips made in a year}} = \dfrac{9300}{1161} = 8.01$ days

- Average service rate $\mu = 0.1248$ per day

5.4.2 Model Execution

Solving such waiting line model manually is a tedious task. The aforementioned queuing model was solved with the help of a computer using queuing theory software package. There are several software packages available for this purpose, and one may refer to any Queuing theory or Operations Research textbook (some of which are listed in "References/Further Reading" at the end of this chapter) to choose an appropriate one. Different software packages require different levels of expertise, and their purpose may vary depending on specificities. A comprehensive list of queuing theory software and related information is available in Ref. [3].

5.5 RESULTS

The various queuing characteristics such as average queue length (L_q) and system length (L), average waiting time in the queue (W_q) and that in the system (W), probability that all servers are idle (P_0), probability that a call waits (P_w), probability of "n" calls in the system (P_n), and server utilization (ρ) have been computed with varying number of OSVs. The results have been summarized in Appendix 5.A.

Since $\lambda/\mu = 25.45$, different queuing characteristics were computed starting with 26 OSVs. This, in other words, represents the first value after the condition $(\lambda/c\mu < 1)$ holds true:

(a) Figure 5.2 exhibits waiting time in the queue (W_q) and that in the system (W), as well as server utilization (ρ) against varying OSVs. It may be noted from Figure 5.2 and Appendix 5.A that waiting time W_q, W declines rapidly with the increase of OSVs and stabilizes at a low value $(W_q = 0.31, W = 8.32)$ corresponding to OSV=31. Thereafter, the reduction of W and W_q is marginal with the increase in OSV. The server utilization is also high (82.2%) corresponding to OSV = 31.

The optimum number of OSVs for a specified service level can also be determined from Appendix 5.A and Figure 5.2. For example, server utilization is 91% with 28 OSVs, 88% with 29 OSVs, 85% with 30 OSVs, 82.2% with 31 OSVs, and so on.

A company needs to fix its acceptable service level based on optimization of total system cost. In order to optimize the total cost, it is required to strike a balance between the waiting cost of installations and the service cost of OSVs. *The process and steps to be followed for optimization of total system cost have been elaborated in* Section 5.6e. However, relevant cost data of various installations and servers (OSVs) were not sufficiently available. Therefore, detailed cost optimization could not be incorporated in this study. Nonetheless, empirical observations suggest that *waiting cost of installations is about five times higher than the service cost (i.e., operating cost of OSV). Therefore, trade-off between capacity increase (additional OSVs) and reduction in waiting time of installations would lead to optimization of total system cost.*

Keeping this in mind, it may be seen from Appendix 5.A and Figure 5.2 that waiting time in the queue (W_q) decreases with the increase in OSVs as follows: $W_q - 1.675$ days for

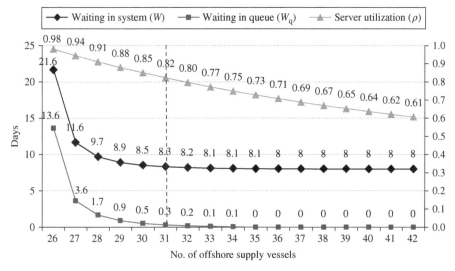

FIGURE 5.2 Queuing characteristics: waiting time (W, W_q) versus the number of OSVs.

28 OSVs, 0.9 days for 29 OSVs, 0.522 days for 30 OSVs, and 0.311 days for 31 OSVs. In other words, *the increase in OSVs from 28 to 31 (addition of 3 OSVs) would reduce the waiting time of installations to one-fifth or less (from 1.675 days to 0.311 days). Therefore, 31 OSVs are found to be optimum for the current activity level.*

(b) Figure 5.3 depicts the characteristics of average queue length (L_q) and system length (L), as well as server utilization (ρ) with varying OSVs. It may be seen from Figure 5.3 and Appendix 5.A that the average entities in the queue (L_q) and in the system (L) also follow similar trend and decline steeply with the rise in OSVs. Both L and L_q stabilize at a low value $(L_q = 0.99, L = 26.47)$ corresponding to OSV = 31, beyond which reduction in L and L_q is minimal with additions of OSVs.

It may also be seen that queue length (L_q) corresponding to 28 OSVs is 5.32, which reduces to 2.87 for 29 OSVs, 1.66 for 30 OSVs, and 0.99 for 31 OSVs, that is, *the increase in service capacity (addition of 3 OSVs) decreases queue length to one-fifth or less (from 5.32 to 0.99). This conforms to the earlier findings that 31 OSVs are optimum at the current activity level.*

(c) The other operating characteristics, namely, the probability that a call waits (P_w), the probability of n calls in the system (P_n), and the probability that all servers are idle (P_0) against varying OSVs, are shown in Figure 5.4.

It may be seen from Figure 5.4 and Appendix 5.A that P_w, P_n, and P_0 change rapidly with the increase in OSV until they stabilize at OSV = 31. Thereafter, changes are not appreciable with the addition of OSVs. It may also be noted that the value of P_0 is in the order of 10^{-11}; therefore, the occurrence that all servers would be idle is quite unlikely. *All these conform that 31 OSVs are optimum and sufficient.*

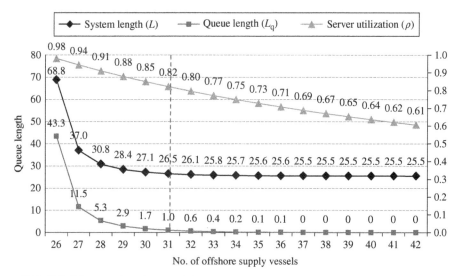

FIGURE 5.3 Queuing characteristics: queue length (L, L_q) versus the number of OSVs.

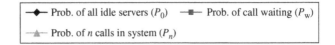

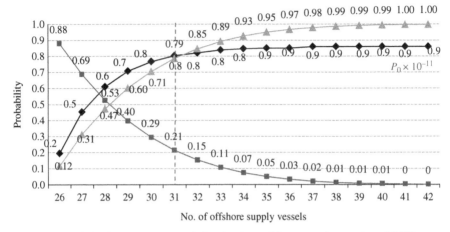

FIGURE 5.4   Queuing characteristics: P_0, P_w, and P_n versus the number of OSVs.

A sample of the final solution for $M/M/31$ model is shown in Appendix 5.B.

5.5.1   Recommendation

The optimum OSV fleet size has been determined using queuing theory with a view to optimize waiting time of installations, which in turn would lead to optimization of system cost. The detailed results of the queuing model have been explained in Sections 5.5a–c. A company may fix its desired (optimum) service level accordingly. Based on empirical cost data, it is found that the average waiting cost of offshore installations is about five times higher than the average servicing cost (i.e., average operating cost of OSVs). Therefore, it is desirable that offshore installations waiting for OSVs be minimized. Taking this as prime criteria (i.e., minimize W_q and W) along with others as mentioned in Sections 5.5a–c, it is found that *31 OSVs are sufficient and appropriate for the current activity level*. Other performance measures such as L, L_q, P_w, P_n, and P_0 also conform this.

Therefore, *it is recommended that 31 OSVs be deployed at the current activity level, which would optimize waiting time of installations*.

5.6   LIMITATIONS OF THE MODEL AND SCOPE FOR FURTHER WORK

Queuing theory is essentially a rigorous mathematical treatise and is not easy to master the subject. It requires mathematical knowledge and analytical skills to understand, apply, and solve such problems. It is generally found that practicing managers, professionals, and others shy away from mathematical equations and derivations. This dampens their interest in using quantitative tools and techniques such as queuing theory. The limitation of the current model and scope for further work are mentioned in the following paragraphs:

(a) The model assumes steady-state condition, as E&P and OSV operations are going on for long and have sufficiently settled down. Queuing models are mostly based on steady-state condition, and for all practical purposes, they provide the desired or satisfactory results. Nonetheless, critics may question and ask for test-ing the model in transient condition, which is far more complex. Therefore, further work on developing/testing the model in transient condition may be explored.

In this context, it is worth mentioning that a complex model not necessarily means a powerful model or offers elegant solutions. In reality, some waiting line situation may be complex. But complicated mathematical model does not neces-sarily offer the best economic solution. On the other hand, relatively simple and approximated model often provides acceptable result satisfying both qualitative and quantitative aspects of the complex problem.

(b) The model assumes interarrival time and service time as exponential (Markovian) distribution, which is based on the pilot study/testing of call rate data and service rate data. This can be further refined by elaborately testing the call rate and service rate data, so that there is no room for doubting the robustness of the model. For this purpose, analyze interarrival times (say, $t_1, t_2, t_3, \ldots t_n$) and a reasonable estimate of exponential interarrival rate λ is given by $\lambda = n / \sum_{i=1}^{i=n} t_i$.

(c) It is important to consider the desired service level while designing the queuing system. In other words, the service level acceptable to the company or the company intends to achieve is to be firmed up for designing the queuing system. This exercise was carried out assuming that the current utilization of OSVs is accept-able to the company. Apart from this, further studies may be carried out to explore possibilities of improvement in OSV utilization through better maintenance plan, logistic or deployment plan, space utilization, and so on.

(d) Queuing model essentially tries to balance the "cost of waiting" and "cost of service" with the objective of minimizing total system cost. A typical waiting cost and service cost or capacity trade-off model is depicted in Figure 5.5:

- Minimize total system cost (TC) = waiting cost (WC) + service cost (SC).
- Waiting cost (WC) = $C_w L_q$ (where C_w is the cost of waiting per customer per unit time and L_q is the number of customers in the queue).
- Service cost (SC) = $c * C_s(\mu)$ (where c is the number of servers, μ is the average service time per customer, and C_s is the cost per server per unit time as a function of μ).
- Min TC = $C_w L_q + c * C_s(\mu)$.
- Optimize by enumerating various combinations of μ and c, compute the total cost (TC) for each case, and choose the cheapest alternative.

(e) The cost of additional capacity (increase in the number of OSVs) must be bal-anced against the reduction in cost on account of waiting for OSVs. The waiting

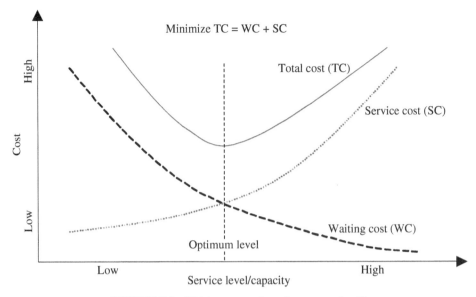

FIGURE 5.5   Waiting cost and service cost trade-off.

cost of each installation (rigs, platforms, etc.) and that of each OSV may be collected for optimization of cost. Such cost data were not sufficiently available while carrying out this study; therefore, these aspects could not be included in detail in this study. This would quantify the monetary impact of total system cost and opportunity for trade-off between waiting cost and service cost. Further work in these areas may be carried out in the future.

5.7   CONCLUSION

The waiting cost of offshore installations such as drilling rigs and production platforms is exorbitantly high. It is desirable that no offshore installation waits for OSVs that supplies materials. Therefore, adequate numbers of OSVs are necessary for timely delivery of materials to numerous installations and maintain uninterrupted E&P operations. But OSVs too are costly items, whose average operating cost is nearly 20% of average offshore rig operating cost. Therefore, it is necessary to trade off between the waiting time of installations and the service cost of OSVs.

The current study estimates optimum OSV fleet size using queuing theory with the objective of optimizing waiting time of installations (thereby optimizing overall system cost). It also determines various performance measures, such as waiting time in the queue and that in the system, queue length and system length, servers' utilization, probability of call waiting, probability of specified calls in the system, and probability of all servers being idle, which would be useful for decision making under varying circumstances. *The model suggests that 31 OSVs are sufficient and appropriate at the current level of activity, which would optimize waiting time of installations.*

Benefits: There is considerable scope for saving by optimizing OSV fleet size. Considering that the average operating cost of an OSV is approximately USD 15,000 per day, there would be a saving of USD 5.5 million per year on account of reduction of one OSV. The saving potential will multiply with the reduction in OSVs by optimizing fleet size. Apart from this, improvement in utilization of OSVs through various means (such as better logistic planning and scheduling, regular and scheduled OSV maintenance, capacity utilization in terms of floor space, weight and loading, etc.) would also help in optimizing OSV fleet size.

Chapter 6 aims to standardize consumption of HSD, cement, and chemicals in oil/gas wells and rigs, which vary widely due to various geological, technical, and physical factors. There is ample scope for improvement and standardizing consumption of these consumables, which would help in preparing reliable procurement plan, avoiding excess inventory, and preventing stock-out situation.

REVIEW EXERCISES

5.1 What are the jobs normally done by OSVs? Why it is important to optimize OSV fleet size?

5.2 Give some examples where queuing theory can be applicable related to your industry or domain of work.

5.3 As a queuing theory practitioner, what are the steps you would suggest to formulate and solve a waiting line problem?

5.4 What are the basic components of a queuing system? Describe them.

5.5 What are the operating characteristics of a queuing model? Explain their importance.

5.6 What do you understand by the following terms?

Birth–death process
Markovian chain
Poisson arrival
exponential service
traffic intensity
transient and steady-state condition
balking
jockeying
reneging
Kendall's notation
M/M/c model

APPENDIX 5.A QUEUING CHARACTERISTICS WITH VARYING NUMBER OF OFFSHORE SUPPLY VESSELS

No. of OSVs	System Length (L)	Queue Length (L_q)	Waiting Time in System (W), Days	Waiting Time in Queue (W_q), Days	Prob. of All Idle Servers ($P_0 \times 10^{-11}$)	Prob. of Call Waiting (P_w)	Prob. of n Calls in System (P_n)	Server Utilization (ρ)
26	68.81	43.330	21.64	13.620	0.195	0.880	0.117	0.980
27	37.00	11.520	11.637	3.620	0.453	0.687	0.313	0.944
28	30.80	5.320	9.688	1.675	0.61	0.526	0.473	0.910
29	28.35	2.870	8.91	0.904	0.707	0.397	0.602	0.879
30	27.14	1.660	8.53	0.522	0.766	0.294	0.705	0.850
31	26.47	0.990	8.32	0.311	0.802	0.214	0.786	0.822
32	26.08	0.600	8.2	0.188	0.82	0.153	0.846	0.796
33	25.84	0.360	8.12	0.114	0.838	0.107	0.892	0.772
34	25.70	0.220	8.08	0.069	0.846	0.074	0.926	0.749
35	25.61	0.130	8.05	0.042	0.85	0.050	0.950	0.728
36	25.56	0.080	8.038	0.025	0.85	0.033	0.967	0.708
37	25.52	0.047	8.027	0.015	0.86	0.021	0.979	0.688
38	25.50	0.028	8.021	0.009	0.86	0.013	0.986	0.670
39	25.49	0.016	8.017	0.005	0.86	0.008	0.992	0.653
40	25.49	0.009	8.015	0.003	0.86	0.005	0.995	0.637
41	25.49	0.005	8.014	0.002	0.86	0.003	0.997	0.622
42	25.48	0.003	8.013	0.001	0.86	0.001	0.998	0.607

APPENDIX 5.B   FINAL SOLUTION FOR *M/M/*31 MODEL

With $\lambda = 3.18$ calls per day and $\mu = 0.1248$ customers per day
Overall system effective arrival rate = 3.179999 per day
Overall system effective service rate = 3.179999 per day
Overall server effective utilization factor = 0.821960
Average number of entities in the system $(L) = 26.4713$
Average number of entities in the queue $(L_q) = 0.990481$
Average time spent in the system $(W) = 8.324296$ days
Average time an entity waits in the queue $(W_q) = 0.311472$ days
The probability that all servers are idle $(P_0) = 0.80E{-}11$
The probability an arriving entity waits $(P_w) = 0.214542$

Probability of "*n*" Entities in the System

$P(0) = 0.00000$	$P(1) = 0.00000$	$P(2) = 0.00000$	$P(3) = 0.00000$
$P(4) = 0.00000$	$P(5) = 0.00000$	$P(6) = 0.00000$	$P(7) = 0.00001$
$P(8) = 0.00004$	$P(9) = 0.00010$	$P(10) = 0.00026$	$P(11) = 0.00059$
$P(12) = 0.00126$	$P(13) = 0.00246$	$P(14) = 0.00448$	$P(15) = 0.00761$
$P(16) = 0.01211$	$P(17) = 0.01816$	$P(18) = 0.02570$	$P(19) = 0.03447$
$P(20) = 0.04391$	$P(21) = 0.05328$	$P(22) = 0.06171$	$P(23) = 0.06837$
$P(24) = 0.07259$	$P(25) = 0.07398$	$P(26) = 0.07251$	$P(27) = 0.06843$
$P(28) = 0.06227$	$P(29) = 0.05471$	$P(30) = 0.04647$	

$$\sum_{i=0}^{30} P_{(i)} = 0.785458$$

REFERENCES

Useful Links

[1] BIS 494: Topics in Information System, Introduction to Queuing and Simulation, in Business Process Modelling, Simulation and Design, available at: www.nku.edu/~sakaguch/ifs494/CH06.PPT (accessed on March 19, 2016).

[2] Newman, R., Elementary Queuing Theory Notes, available at: http://www.cise.ufl.edu/~nemo/cop5615/queuing.html (accessed on March 4, 2016).

[3] List of Queuing Theory Software, compiled by Dr. Myron Hlynka, University of Windsor, Canada, available at: http://web2.uwindsor.ca/math/hlynka/qsoft.html (accessed on March 4, 2016).

FURTHER READING

Allen, A.O., Probability, Statistics, and Queuing Theory with Computer Science Applications, 2nd edition, Academic Press, Inc., Boston, MA, 1990.

Benes, V.E., General Stochastic Processes in the Theory of Queues, Addison Wesley Pub. Co., Boston, MA, 1963.

Bhat, U.N., An Introduction to Queueing Theory: Modeling and Analysis in Applications, Birkhäuser, Boston, MA, 2008.

Bose, S.K., An Introduction to Queuing Systems, Kluwer Academic/Plenum Publishers, New York, 2002.

Chee-Hock, N. and Boon-He, S., Queuing Modeling Fundamentals, 2nd edition, John Wiley & Sons, Ltd, Chichester, 2002.

Cooper, R.B., Introduction to Queuing Theory, 3rd edition, CEE Press, Washington, DC, 1990.

Dshalalow, J.H., Advances in Queueing: Theory, Methods, and Open Problems, 1st edition, CRC Press, Boca Raton, FL, 1995.

Ganesh, A., O'Connell, N., and Wischik, D.J., Big Queues, Lecture Notes in Mathematics, Springer, Berlin, 2010.

Gnedenko, B. and Kovalenko, I., Introduction to Queuing Theory, Birkhäuser, Boston, MA, 1991.

Gross, D., Shortle, J.F., Thompson, J.M., and Harris C.M., Fundamentals of Queuing Theory, 4th edition, John Wiley & Sons Inc., New York, 2008.

Kleinrock, L., Queuing Systems. Vol. I. Theory. John Wiley & Sons, New York, 1975.

Stewart, W., Probability, Markov Chains, Queues, and Simulation, Princeton University Press, Princeton, NJ, 2009.

Sztrik, J., An Introduction to Queuing Theory and Its Applications (in Hungarian), Kossuth Egyetemi Kiadó, Debrecen, 2000.

Sztrik, J., Basic Queuing Theory, University of Debrecen, Debrecen, 2012, available at: http://irh.inf.unideb.hu/user/jsztrik/education/16/SOR_Main_Angol.pdf (accessed on March 4, 2016).

Takács, L., Introduction to the Theory of Queues, Oxford University Press, New York, 1962.

Wagner, H.M., Principles of Operations Research, Prentice-Hall of India, New Delhi, 1980.

Useful Link

Queuing Theory Presentation.ppt – Isites Harvard Edu, available at: http://isites.harvard.edu/fs/docs/icb.topic1115859.files/Queuing%20Theory%20Presentation.ppt (accessed on March 19, 2016). As part of Vaccaro, P.A., Applied Management Science for Decision Making, 1st edition, Pearson Prentice-Hall, Inc., 2012.

6

STANDARDIZING CONSUMPTION OF HSD, CEMENT, AND CHEMICALS IN OIL/GAS WELLS AND RIGS

6.1 INTRODUCTION

In previous chapters, we have seen that oil/gas wells are capital intensive and require specialized operations, equipment, and materials. In this chapter, we would deal with consumables used in drilling and completion of oil and gas wells. Typical consumables used in drilling oil and gas wells are POL (high-speed diesel (HSD) and lube oil), cement, chemicals, spares, and other miscellaneous items, which cost substantial amount. This study focuses on major consumables that are associated with critical operations and are required in bulk quantity such as HSD, oil well cement, and chemicals. These materials are vital for performing important operations such as running engines, cementing jobs, and preparing drilling fluids.

HSD is used in drilling rig as fuel for generating power for carrying out all operations round the clock. Usually, DG sets are used to generate power in drilling rigs. Specially designed oil well cement is used for cementing jobs such as casing, squeezing, plugging, and so on. Good cementing jobs are crucial for the success of drilling and well completion. Chemicals are used in preparing drilling fluids, which is considered as "lifeblood" of drilling operation. These three consumables—namely, HSD, cement, and chemicals—are important ingredients for completing oil and gas wells.

The consumption of HSD, cement, and chemicals for different wells/rigs varies due to various geological, technical, and physical reasons. The wide variability of factors influencing consumption of these materials makes it difficult to standardize. This study aims to standardize consumption of HSD, cement, and chemicals in oil/gas well and rig. It discusses on the need and scope for standardizing consumption and its benefits, factors influencing consumption, and reasons for variations in consumption and suggests

Optimization and Business Improvement Studies in Upstream Oil and Gas Industry, First Edition.
Sanjib Chowdhury.

measures for improvement. All these have been elaborated in separate sections in this chapter (HSD, Section 6.4; cement, Section 6.5; and chemicals, Section 6.6).

6.2 OBJECTIVES

Standardizing consumption of consumables in oil/gas wells and rigs is a complex issue, as there are several factors influencing consumption of each consumable. Also, these factors vary widely across different geological, technical, and physical conditions. The objective of the current study is as follows:

(a) To standardize the consumption of HSD, cement, and chemicals in oil/gas wells and rigs. This, in turn, would help in meeting the following subobjectives:
   (i) To enable planners to work out future consumption with fair degree of accuracy and plan procurement accordingly
   (ii) To prevent stock-out situations, so that rigs do not wait for want of materials
   (iii) To avoid excess inventory that blocks capital and warehouse space
   (iv) To monitor consumption and check wastage and aberrations

This study was conducted in an E&P company having multiple Assets and Basins spread over a vast geographical areas in onshore and offshore. Onshore Assets and Basins depending on their geographical proximity are grouped together within region for administrative purpose and convenience. Assets/Basins in the same region usually have similar geologic conditions. The company has three major onshore regions—namely, Region 1, Region 2, and Region 3—as explained in Section 6.3d.

6.3 METHODOLOGY

The methodologies followed for this study are described as follows:

(a) HSD consumption data of own and contract rigs operating in onshore and offshore was collected for 3 years (Year 1, Year 2, and Year 3) and analyzed.
(b) Cement and chemical consumption data of approximately 330 wells drilled in onshore and offshore for 1–1½ years were collected and studied.
(c) Parameters for fixing standards for HSD, oil well cement, and chemicals were carefully chosen considering the relevant factors affecting these.
   • Rigs of similar type and capacity operating in similar geological formation and depth (having similar ROP) for onshore and offshore areas were grouped for fixing standards for HSD consumption.
   • Likewise, wells of similar casing policy, hole size, and depth in the same field/region were grouped together for fixing standards for oil well cement. Wells having a deviation of ±20% from the average depth have been excluded from this study (being outliers).
   • Similarly, unit consumption of chemicals in similar type of wells in a field/Asset/Basin was considered for standardization of consumption of chemicals.

(d) The company has the legacy of grouping onshore Assets/Basins based on their geographical location; thus, Assets/Basins located in the same part of the country or geographical area are called "region." There are three major onshore regions that contain majority of the Assets/Basins. *Assets/Basins within the same region are generally found to be of similar nature in terms of subsurface characteristics and geology, besides having cultural homogeneity and geographical closeness.*

Thus, Region 1 comprises Asset 1A, Asset 1C, Asset 1M, Asset 1N, and so on. Similarly, Region 2 contains Asset 2A, Asset 2T, Basin 2AA, Basin 2B, Basin 2C, and so on. Asset 3K, Asset 3R, and others are in Region 3.

The process and methodology of fixing standards for the consumption of HSD, oil well cement, and chemicals have been further elaborated in the respective sections.

6.4 STANDARDIZATION OF HSD CONSUMPTION

Deep drilling rigs are equipped with high-powered engines for generating electrical power to drive equipment and tools and carry out all drilling and nondrilling operations round the clock at the drill site. Diesel engine (DG set) is the most commonly used power generator, which uses HSD as fuel. POL used in drilling rigs accounts for approximately 6.3% of rig operating cost or approximately 3.8% of well cost (Section 2.4g) and mainly consists of HSD and lube oil. HSD accounts for nearly 99% of POL consumed and lube oil remaining 1%. Therefore, standardization has been confined to the consumption of HSD.

6.4.1 Choosing Parameters for Fixing Standards

The consumption of fuel in a deep drilling rig depends on several factors, such as the following:

(i) Type and capacity of rig
(ii) Type of operations—drilling or workover
(iii) Phases of operation—drilling, production testing, repairs, rig move, and so on.
(iv) Depth and type of formation being drilled, rate of penetration (ROP), well complications, and so on.
(v) Efficiency, age, and condition of engines
(vi) Efficiency, age, and condition of different equipment such as mud pumps, draw works, compressors, motors, and so on.

Engine efficiency is one of the factors influencing fuel consumption in drilling rigs. Fuel (HSD) consumption of engine is dependent on engine size and load on engine. Engine efficiency is generally expressed in terms of power generated per unit of fuel consumption at a specified load and is measured in "kWh/l." Optimum fuel consumption (chart in kWh/l) for different engine size and varying loads are generally specified by original equipment manufacturers (OEMs). Engine efficiency is a part of overall drilling performance or efficiency, which includes many other parameters.

Fuel efficiency of engine may be same or comparable for productive drilling and nonproductive drilling operations. For example, the faster the drilling, the higher the fuel consumption, as the latter rises with the increase in torque, revolutions per minute (RPM), pump output and pressure, and so on; rigs with prolonged downhole complications, repair, and production testing consume less fuel and may have fuel consumption efficiency (kWh/l) similar to productive drilling operation.

Footage or meterage drilled is the primary focus of drilling operation. Faster drilling conforming to standard operating practice saves precious rig time, which outweighs saving on account of fuel consumption, as the latter is a small part compared to rig hiring cost. On a broader level, overall fuel and drilling efficiency of rigs being the primary concern, "fuel consumption per meter or foot drilling" has been chosen as a measure of fuel efficiency of drilling rig, which takes into account all the aforementioned factors (i–vi). This measure is in conformity with other standard measures of drilling efficiency used in the E&P industry such as drilling cost per meter or foot, meterage or footage drilled per day, and so on. For the purpose of parity, similar type and capacity rigs operating in similar geological formation and depth were grouped for standardizing fuel consumption of drilling rigs for both on land and offshore.

The company under reference uses "fuel consumption per rig-month" as standard for measuring fuel efficiency in drilling rig. Such standard does not indicate true measure of fuel efficiency, which is evident from the fact that there is wide disparity in "fuel consumption per meter or foot drilling" between own rigs and hired rigs as will be shown later in this chapter.

6.4.2 Fixing Standards for HSD Consumption

HSD consumption data of onshore drilling rigs for 2 years (Year 1, Year 2) and offshore rigs for 3 years (Year 1, Year 2, Year 3) were analyzed:

(a) Similar type and capacity of drilling rigs operating in similar geological formation and depth were grouped together for fixing standard. Thus fuel consumption standards for various types and capacity of drilling rigs operating in different regions/Assets/Basins were computed.

(b) The fuel consumption per meter drilling by each rig was computed and compared over a period of 2–3 years. It is presumed that each rig over a period of 2–3 years would undergo different phases of operation and would even out skewed consumption, if any, during a phase or activity.

(c) Abnormally high unit fuel consumption values observed in certain year by certain rigs were excluded while computing average unit consumption and standard deviation of each category of rigs. These high values are mainly due to exceptional circumstances, such as prolonged downhole complications, production testing, force majeure, shallow wells drilled by high-capacity rigs due to nonavailability of appropriate type/capacity of rig, and other reasons leading to low meterage/footage drilled.

(d) There are some rigs that are of different types and capacities, or are few in numbers, or used for a specific depth (shallow or deep) in a region. In such cases, the data available for standardizing fuel consumption are insufficient and could not be standardized for few such drilling rigs operating in different regions.

6.4.2.1 Onshore: The average HSD consumption per meter drilling (μ) and standard deviation (σ) for each category of rigs operating in different regions (Assets/Basins) have been computed and are summarized as follows:

(a) Region 1 (Asset 1A, Asset 1M, Asset 1N, and Asset 1C)
  (i) **RY-700s** (700 HP, drill depth up to 2100 m): $\mu = 29$ l/m, $\sigma = 2.9$, and $N = 79$ (refer to Fig. 6.1; N, number of wells)
  (ii) **RY-700s** (700 HP; rated drill depth, 3000 m): $\mu = 44$ l/m, $\sigma = 3.7$, and $N = 61$ (refer to Fig. 6.2; N, number of wells)
  (iii) **RX-1000s** (1000 HP; rated drill depth, 3600 m): $\mu = 75$ l/m, $\sigma = 15$, and $N = 36$ (refer to Fig. 6.3; N, number of wells)
  (iv) **RX-1400s** (1400 HP; rated drill depth, 4900 m): $\mu = 86$ l/m, $\sigma = 12$, and $N = 14$ (refer to Fig. 6.4; N, number of wells)
     Y-axis gridlines in the following figures depict μ, $\pm\sigma$, and $\pm 2\sigma$ values and ranges. The X-axis denotes rig names, capacity, and Asset name(s).

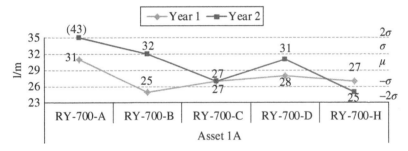

FIGURE 6.1 Region 1: HSD consumption per meter drilling; drill depth, 2100 m.

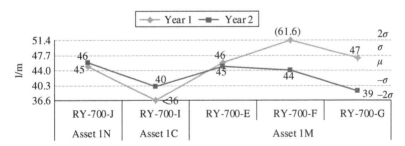

FIGURE 6.2 Region 1: HSD consumption per meter drilling; drill depth, 3000 m.

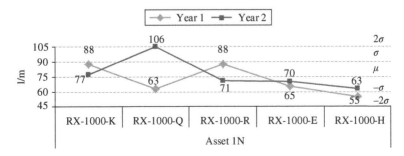

FIGURE 6.3 Region 1: HSD consumption per meter drilling; drill depth, 3600 m.

(b) Region 3 (Asset 3R and Asset 3K)
   (i) **RX-1000s** (1000 HP; rated drill depth, 3600 m): $\mu = 56$ l/m, $\sigma = 10$, and $N = 26$ (refer to Fig. 6.5; N, number of wells)
   (ii) **RX-1400s** (1400 HP; rated drill depth, 4900 m): $\mu = 104$ l/m, $\sigma = 16.7$, and $N = 20$ (refer to Fig. 6.6; N, number of wells)
   (iii) **RZ-2000s, RB-2000s, and RX-2000s** (2000 HP; rated drill depth, 6100 m): $\mu = 118$ l/m, $\sigma = 29$, and $N = 12$ (refer to Fig. 6.7; N, number of wells)
   Y-axis gridlines in the following figures depict μ, $\pm\sigma$, and $\pm2\sigma$ values and ranges. The X-axis denotes rig names, capacity, and Asset name(s).

(c) Region 2 (Asset 2A, Asset 2T, Basin 2AA, Basin 2B, and Basin 2C)
   (i) **RX-1000s** (1000 HP; rated drill depth, 3600 m): $\mu = 131$ l/m, $\sigma = 50$, and $N = 36$ (refer to Fig. 6.8; N, number of wells)

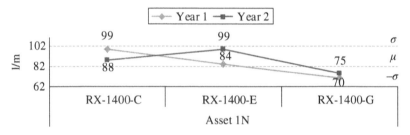

FIGURE 6.4   Region 1: HSD consumption per meter drilling; drill depth, 4900 m.

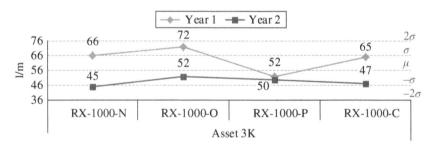

FIGURE 6.5   Region 3: HSD Consumption per meter drilling; drill depth, 3600 m.

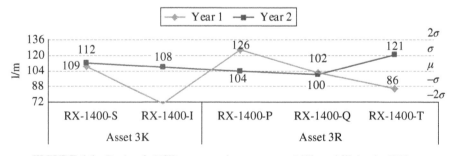

FIGURE 6.6   Region 3: HSD consumption per meter drilling; drill depth, 4900 m.

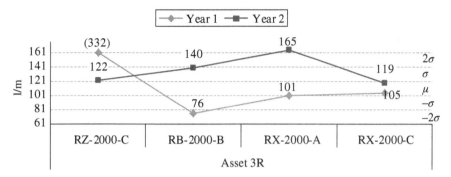

FIGURE 6.7 Region 3: HSD consumption per meter drilling; drilling depth, 6100 m.

(ii) **RX-1400s** (1400 HP; rated drill depth, 4900 m): $\mu = 184$ l/m, $\sigma = 52$, and $N = 42$ (refer to Fig. 6.9; N, number of wells)

(iii) **RB-2000s, RX-2000s, and RA-2000s** (2000 HP; rated drill depth, 6100 m): $\mu = 244$ l/m, $\sigma = 109$, and $N = 22$ (refer to Fig. 6.10; N, number of wells)

The high standard deviation for all categories of rigs in Region 2 denotes the necessity of further analysis with more well data.

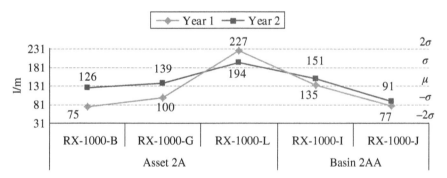

FIGURE 6.8 Region 2: HSD consumption per meter drilling; drill depth, 3600 m.

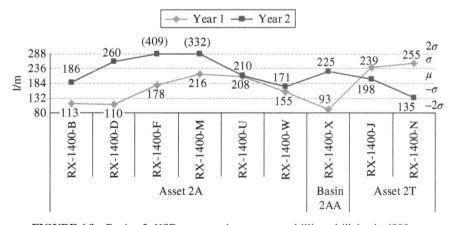

FIGURE 6.9 Region 2: HSD consumption per meter drilling; drill depth, 4900 m.

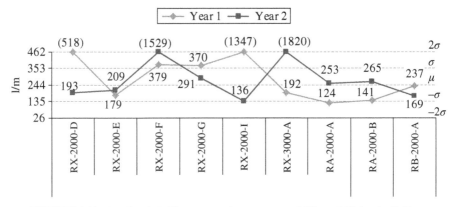

FIGURE 6.10   Region 2: HSD consumption per meter drilling; drill depth, 6100 m.

Y-axis gridlines in the aforementioned figures depict μ, $\pm\sigma$, and $\pm2\sigma$ values and ranges. The *X*-axis denotes rig names, capacity, and Asset/Basin names.

6.4.2.2 Deep Dive and Inferences: The fuel consumption analysis of different capacity rigs operating in various fields/regions reveals many interesting facts, which are mentioned in the following paragraphs:

(a) *It is observed that more than 95% population in each category of rig in Assets/ Basins of Regions 1 and 3 fall within 2σ limit, implying statistical consistency and validity of the result. It is also seen that σ increases with depth as uncertainties involved and variance with respect to downhole complications, well completion, and production testing time increase with depth.*

(b) It is observed that unit fuel consumption of most of the rigs in onshore Region 2 including Asset 2A, Asset 2T, Basin 2AA, and Basin 2B is much higher than that of similar rigs operating at similar depth in other Assets/Basins in Region 1 and Region 3. The reasons for such wide variations are varied, which are elaborated as follows:

   (i) The subsurface characteristics of Region 2 are markedly different from that of Region 1 and Region 3. The formation is relatively hard, complex, and prone to complications in Region 2; the ROP is also low compared to Region 1 and Region 3. All these have considerable influence on HSD consumption; as a result, unit fuel consumption is much higher in Assets/Basins of Region 2 compared to Region 1 and Region 3.

   (ii) *Higher fuel consumption in Region 2 is also due to improper or underutiliza-tion of higher-capacity drilling rigs, where relatively shallow wells are drilled with higher-capacity rigs. On several occasions, rigs are deployed primarily on the basis of their availability and not on their suited capacity.* The wide variation between Region 2 and Region 1 and Region 3 requires further analysis and fine-tuning of rig deployment plan. Deployment of suitable types/capacity rigs would reduce HSD consumption, and optimiza-tion of rig distribution plan may be encouraged.

   (iii) Maintenance of rig engines and motors and load distribution design influence HSD consumption in drilling rigs. Higher fuel consumption in Region 2 also indicates poor maintenance of rig equipment, engines, and so on. In-depth study of periodic maintenance schedule would help in improving fuel consumption in drilling rigs.

(iv) Apart from geological and technical reasons, pilferage, malpractices, and so on, are also contributing factors for high fuel consumption in Region 2. It is alleged that pilferage of HSD is a well-organized racket in few Assets/Basins in Region 2, which are in remote, trouble-torn areas with serious law and order problem. So fixing standard for HSD consumption may act as a monitoring tool for checking aberrations.

6.4.2.3 Offshore: There are 10 owned offshore rigs—two drillships and eight jack-up rigs. Fuel consumptions for drilling operations have been considered for computing unit fuel consumption of drilling rigs based on 3 years' data (Year 1, Year 2, and Year 3).

Similar type and capacity rigs operating in similar geological formation or same field were grouped together as follows:

1. Own jack-up rigs for drilling operation
2. Own drillships for drilling operation

Fuel consumption of own rigs as grouped earlier was compared with that of similar type, capacity, and groups of hired rigs:

3. Hired jack-up rigs for drilling operation
4. Hired drillships for drilling operation

The rated capacity and other details of own and hired rigs are shown in Appendix 6.A:

(a) It may be seen from Appendix 6.A that own jack-up rigs can be compared with most of the hired jack-up rigs for the purpose of fuel consumption, as they are of similar type and capacity, operating in the same field at similar depth.

(b) HSD consumption cannot be compared for drillships (including own and hired) due to wide variation in capacity and applications.

(c) *The average HSD consumption per meter drilling is 80% higher in own jack-up rigs (314 l/m) than that of hired rigs (173 l/m). The average fuel consumption per meter drilling by hired jack-up rigs may be followed as standard for own jack-up rigs (Fig. 6.11).*

(d) HSD consumption per meter drilling (μ) and standard deviation (σ) of own as well as hired jack-up rigs are given in the following table and graphically shown in Figures 6.12 and 6.13 for reference.

Jack-up rigs	Consumption
Own (jack-up #1, 2, 3, 4, 5, 6)	$\mu = 314$ l/m
	$\sigma = 66, N = 48$
Hired (jack-up #1, 2, 3, 4, 5, 6, 7, 8, 9, 10)	$\mu = 173$ l/m
	$\sigma = 43, N = 91$

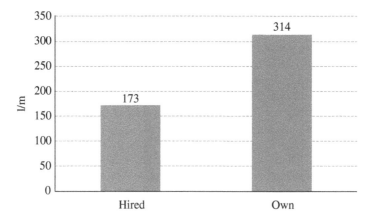

FIGURE 6.11 Comparison of HSD consumption per meter drilling.

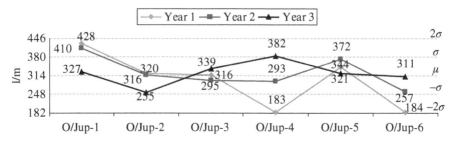

FIGURE 6.12 HSD consumption per meter drilling by own jack-up rigs.

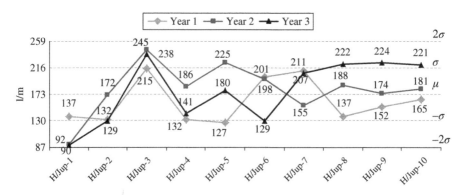

FIGURE 6.13 HSD consumption per meter drilling by hired jack-up rigs.

Y-axis gridlines in the aforementioned figures depict μ, $\pm\sigma$, and $\pm2\sigma$ values and ranges. The X-axis denotes rig names and types.

(e) The wide difference in unit fuel consumption between own and hired offshore drilling rigs is mainly due to operational efficiency. This is amply clear as hired offshore rigs complete a well much faster than own rigs due to full rig time utilization, improved technology, better work practices, and professional culture. There is a lot of scope for improvement of own rigs and associated crews in these areas.

(f) The unit fuel consumption of drillships **cannot be compared as they are of different capacities**. HSD consumption per meter drilling (μ) of own as well as hired drillships as observed is given in the following table and depicted in Figures 6.14 and 6.15.

Drillships	Consumption (l/m)
Own	
Drillship 1	$\mu = 538$
Drillship 2	$\mu = 358$
Hired	
Drillship 1	$\mu = 463$
Drillship 2	$\mu = 484$

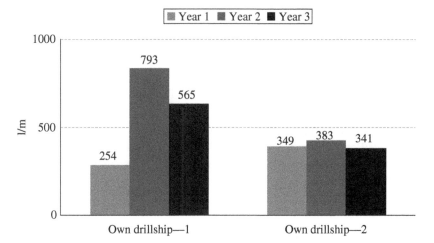

FIGURE 6.14 HSD consumption per meter drilling by own drillships.

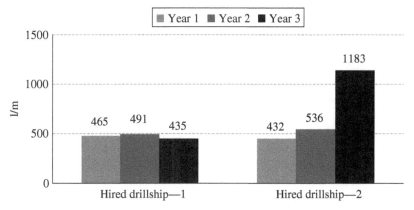

FIGURE 6.15 HSD consumption per meter drilling by hired drillships.

6.4.3 Validation of Result (Standard)

The computed standards/results have been validated and the following inferences are made:

(a) It is observed that computed average HSD consumption (μ) for about 70% of drilling rigs in each category falls within the range of $\mu \pm \sigma$ and that of 95% rigs falls within $\mu \pm 2\sigma$ limit, which is in conformity to normal distribution and validates statistical consistency. This holds true for on-land drilling rigs in Assets/ Basins of Regions 1 and 3 and offshore (jack-up) rigs for drilling operation.

(b) The standard deviation (σ) is found to be very high in all categories of rigs at different drilling depths with wide variations in average consumption (μ) in Region 2. This denotes that further analysis and more well data on fuel consumption are required for fixing fuel consumption standard in Region 2.

(c) It is also seen that σ (variability) increases with depth, which is in conformity with the fact that downhole complications and complexities in drilling rise with depth.

6.4.4 Suggestions

"HSD consumption per meter drilling" is a better indicator and measure of fuel efficiency and may be adopted for standardization, instead of "consumption per rig-month" currently followed in the company, as it embodies rig type and capacity, depth and formation type, ROP, and fuel efficiency of engines and equipment.

"HSD consumption per meter drilling" of each category of rigs operating in different regions/Assets/Basins have been determined in Sections 6.4.2.1a–c for on land and in Section 6.4.2.3d for offshore, which can be specified as standard and used for planning and monitoring purposes:

(a) *Onshore*
   Region 1
      (i) **RY-700s** (700 HP; drill depth up to 2100 m): $\mu = 29$ l/m, $\sigma = 2.9$
     (ii) **RY-700s** (700 HP; rated drill depth, 3000 m): $\mu = 44$ l/m, $\sigma = 3.7$
    (iii) **RX-1000s** (1000 HP; rated drill depth, 3600 m): $\mu = 75$ l/m, $\sigma = 15$
    (iv) **RX-1400s** (1400 HP; rated drill depth, 4900 m): $\mu = 86$ l/m, $\sigma = 12$
   Region 3
     (v) **RX-1000s** (1000 HP; rated drill depth, 3600 m): $\mu = 56$ l/m, $\sigma = 10$
    (vi) **RX-1400s** (1400 HP; rated drill depth, 4900 m): $\mu = 104$ l/m, $\sigma = 16.7$
   (vii) **RZ-2000s, RB-2000s, and RX-2000s** (2000 HP; rated drill depth, 6100 m): $\mu = 118$ l/m, $\sigma = 29$
   Region 2
   (viii) There is a wide variation in average unit fuel consumption (μ) with high standard deviation (σ) for all categories of rigs operating at various drilling depths in Region 2, implying that further analysis delving into more well data is required for fixing unit fuel consumption standard in Region 2.
    (ix) Rigs are often deployed based on their availability. Deployment of higher-capacity drilling rigs in shallow wells results in higher fuel consumption and underutilization of rigs, which may be minimized by optimizing rig

distribution plan based on the work program and inventory of drilling rigs in various Assets/Basins/regions across the company.

(x) The maintenance of rig equipment, engines, motors, and so on, needs further improvement, which would improve the unit fuel consumption.

(b) *Offshore*

Jack-up rigs $\mu = 173\,l/m$, $\sigma = 43$

6.4.5 Engine Efficiency and Load

Engines are the prime movers of drilling rigs without which no operation can take place or equipment and tools can run at drill site. The maximum power at drilling rig is consumed by mud pumps, draw works, rotary table, top drive system, and other equipment. The load on engine depends on operations, such as drilling, backreaming, tripping, mud circulation, and activities requiring relatively low load such as logging, testing, evaluation, and so on. The performance of engine mainly depends on meeting the "total load demand" of rig and handling "rate of change of load."

6.4.5.1 Observations and Deep Dive: The efficiency and load on engines at drill sites were studied and the following observations are made:

(a) The number of engines required at drill site depends on the capacity of rig and depth of wells. The power requirement of drilling rig has increased with the introduction of horizontal drilling and fracturing. Three engines are generally required to drill 12¼" phase in relatively deep wells. Deep drilling rigs of 1400 HP and above capacity are usually equipped with four high-powered engines, and those equal or below 1000 HP are having three engines. However, it has been observed that on an average three engines are running round the clock for all phases of drilling in own offshore drilling rigs at an average load of 25–45% during drilling operation. Similarly, engines in on-land drilling rigs of the company are running at an average load of 20–45% (refer to Fig. 6.16). This has also been confirmed by a detailed study carried out by the company and an independent external agency covering a period of 4 years.

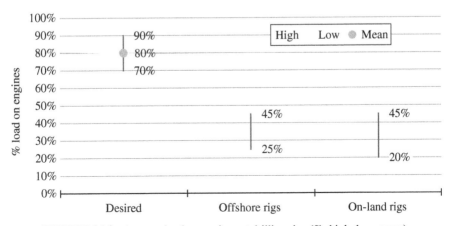

FIGURE 6.16   Average load on engines at drilling rigs (%: high, low, mean).

Diesel engines are generally designed to operate optimally at 0.8 power factor and are most fuel efficient at full power. It is desirable that engines are run at 70–90% load for optimum efficiency and fuel economy (refer to Fig. 6.16). For the instant case, if engines are run at the desired load (70–90%), it would require average running of two engines instead of three engines. Therefore, at least one engine operation can be reduced. In other words, the **life of one engine can be added in each rig**, resulting in a substantial reduction of HSD consumption.

(b) The reduction of one engine operation would ensure greater availability of engines for maintenance and overhauling, thereby increasing their life and possibly reducing the number of standby engines.

(c) The fuel consumption of engine depends on several factors including the health of the engine. Therefore, good maintenance practices play an important role for improving the efficiency of engines.

(d) HSD consumption in rigs may be substantially reduced by ensuring optimal loading of engines. Improved load on engines would lower fuel consumption by an average of 25–30 l/h.

(e) Load monitoring system may be installed at engines of deep-drilling rigs for better load sharing.

(f) There are many old and outdated energy-inefficient diesel engines operating in various drilling rigs of the company, which contributes to the wastage of fuel and energy.

(g) Improvement in operation of mud pumps, agitators, air compressors, and so on, would lead to reduction in fuel consumption.

6.4.6   Suggestions for Improvement

Based on the findings and deep dive in Section 6.4.5, the following suggestions are made, which would minimize energy wastage and contribute toward improvement in fuel consumption standards:

(a) Avoid running more engines simultaneously; instead, improve the average load on engines in drilling rigs from the present level of 25–45% to above 70%. This would reduce the operation of one engine in each rig and ensure greater availability of engines for maintenance and overhauling, which would increase the life of one engine per rig.

(b) For better load management, load monitoring system may be installed at drilling rig engines. Improved load on engines would lower the fuel consumption by an average of 25–30 l/h.

(c) Replace old energy-inefficient engines with new-age energy-efficient engines, which would reduce fuel consumption substantially.

(d) Minimize energy wastage by following standard operating practices of engines, mud pumps, agitators, and air compressors and ensure their optimum usage. For example, during nondrilling operation when load on engine is low, a small DG set may be economical and useful. Similarly, avoid idling of mud pumps and underloading of prime mover, which causes wastage of fuel and energy. Also, ensure proper sizing of agitator system, that is, tank and compartment dimensions, shape, coupling of multiple agitators to single motor, and so on, for optimum use of energy. Likewise, regular maintenance, plugging air leakage, installation of automatic pressure switch, and so on, would considerably reduce wastage of energy in operating air compressors. All these would improve fuel consumption standard.

6.4.7 Benefits

There is a scope of **saving up to USD 62.8 million** on fuel consumption and enhancing life of engines. The savings breakup is as follows:

Offshore
(a) *Fuel consumption cost*: If operational efficiency of own jack-up rigs (314 l/m) is brought to the level of hired rigs (173 l/m), and if it is considered that an average of 33,250 m is drilled by own jack-up rigs in a year (last 2 years' average), and the cost of HSD is at USD 0.9 per liter, then the **saving would be USD 4.2 million per year** (141 l × 33,250 m × USD 0.9 per liter).
(b) *Enhancing life of engines*: If the load on engines is improved from the present 25–45% to 70–90%, then one engine operation can be reduced, or life of one engine will be added to each rig. Considering the average cost of one engine to be USD 0.6 million, the **saving would be USD 6 million** (10 own offshore rigs × USD 0.6 million).

Onshore
(c) *Fuel consumption cost*: If the average load on engine in onshore rigs is improved from the present 20–45% to above 70%, the fuel consumption would reduce by 25–30 l/h, and if it is considered that drilling and production testing phases are around 300 days in a year and cost of HSD is at USD 0.9 per liter, then the **saving would be USD 11.2 million per year** (69 on-land rigs × 24 h × 300 days × 25 l × USD 0.9 per liter).
(d) *Enhancing life of engines*: If the load on engines is improved from the present 20–45% to above 70%, then one engine operation can be reduced, or life of one engine will be added to each rig. Considering the average cost of one engine to be USD 0.6 million, the **saving would be USD 41.4 million** (69 own on-land rigs × USD 0.6 million).

6.5 STANDARDIZATION OF OIL WELL CEMENT CONSUMPTION

This section deals with standardization of cement consumption in oil and gas wells. But before we delve deep, it would be useful to know the basic functions and purpose of cementing operation, besides properties and composition of cementing materials. These are briefly described in the text that follows.

After drilling a certain depth, casing is lowered to prevent the collapse of the borehole wall, and cementing job is done to hold the casing in its position by cement bond. Cement slurry of desired property and density is prepared by mixing cement, additives, and chemicals with water. It is pumped through the pipe and displaced in the annulus between casing and open hole. Oil well cement is basically:

(i) Portland cement that consists of calcium, silica, and aluminum oxide

(ii) Blast furnace slag—a by-product of manufacturing pig iron

(iii) Pozzolans (fly ash—silica/alumina materials)

The American Petroleum Institute (API) has classified oil well cement into nine categories based on its characteristics and use at different range of depths (namely, class A to J and class "G" cement is most commonly used). The properties of cement slurries are changed by adding chemicals and additives, which are known as accelerators, retarders, fluid loss additives, friction reducer, defoamer, and so on.

Cementing operation is necessarily a "one-shot" process with little chance for repeat job; therefore, the first job ought to be the best job. Oil well cementing operation is broadly classified into two categories, namely, (i) primary cementing and (ii) secondary cementing. Primary cementing is essentially cementing of casing and liner with the exposed borehole. All other cementing operations are called "secondary cementing," for example, (i) squeeze cementing and (ii) plug-back cementing.

Primary cementing techniques are mainly of three types, namely, (i) single-stage cementation, (ii) multistage cementation, and (iii) liner cementation. Secondary cementing is done for (i) repairing damages (point leak, split or parted casing), (ii) supplementing faulty primary cementing job, (iii) stopping/controlling lost circulation, (iv) shutting off water, (v) abandonment of well, and so on.

The main functions of cementation are the following:

(i) Provide bonding and support to casing

(ii) Protect casing from corrosion and shock loads

(iii) Seal off problematic zones including high-pressure zone and lost circulation zone

(iv) Restrict or stop fluid movement between formations or in fractured formation

(v) Prevent circulation of mud outside casing

(vi) Close abandoned well

(vii) Sidetrack and so forth

Cementation accounts for approximately 5.6% of well cost and a substantial amount is spent on consumption of cement in a well (Section 2.4g). This study tries to standardize consumption of cement to avoid stock-out situation, so that rigs do not wait for want of cement or cementation jobs.

The consumption of cement in a well depends on the following:

(i) Hole size and depth

(ii) Casing policy—number and sizes of casing

(iii) Cement rise—number of prospective objects to be tested

(iv) Downhole complications—activity/mud loss, caving, and sidetracking

Therefore, *wells with similar casing policy, hole size, and depth in the same field/region have been grouped together for standardization.*

6.5.1 Observations

The cement consumption data of approximately 300 wells drilled during last 1–1½ years in onshore and offshore were analyzed, and the following observations are made:

(a) There is no appreciable difference in the consumption of cement in exploratory and development wells with similar casing policy, hole size, and depth in the same field/region.

(b) In offshore, cement consumption standard for 4 CP and 5 CP wells drilled by jack-up rigs and drillships were attempted to.

(c) About 3–5% wells drilled in a year consume high quantity of cement due to exceptional circumstances such as mud loss, repair jobs, sidetracking, and so on.

6.5.2 Fixing Standards for Consumption of Cement

The average cement consumption (μ) and standard deviation (σ) in wells of similar casing policy, hole size, and depth in different fields/regions have been computed and are furnished as follows:

Onshore
2 CP wells
   (a) **Region 1**: $\mu = 40$ tons, $\sigma = 10$, and $N = 120$ (refer to Fig. 6.17; N, number of wells)
      Other regions do not have 2 CP wells.
3 CP wells
   (b) **Region 1**: $\mu = 69$ tons, $\sigma = 11$, and $N = 27$ (refer to Fig. 6.18; N, number of wells)
   (c) **Region 3** (Asset 3K): $\mu = 69$ tons, $\sigma = 16$, and $N = 13$ (refer to Fig. 6.19; N, number of wells)
   (d) **Region 3** (Asset 3R): $\mu = 112$ tons, $\sigma = 45$, and $N = 20$ (refer to Fig. 6.20; N, number of wells)
      The high standard deviation in consumption of cement in Region 3 implies that more well data are required for fixing standard.
   (e) **Region 2**: $\mu = 88$ tons, $\sigma = 24$, and $N = 32$ (refer to Fig. 6.21; N, number of wells)
4 CP wells
   (f) **Region 1** (Asset 1N): $\mu = 116$ tons, $\sigma = 6$, and $N = 6$ (refer to Fig. 6.22; N, number of wells)
   (g) **Region 2**: $\mu = 189$ tons, $\sigma = 55$, and $N = 8$ (refer to Fig. 6.23; N, number of wells)
      The high standard deviation in Region 2 implies that further well data and analysis are required to fix the standard.
   (h) **Region 3**: Inadequate number of 4 CP wells or data is available for fixing consumption standard.

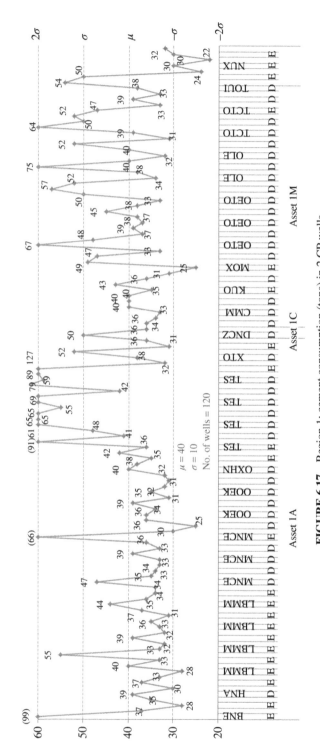

FIGURE 6.17 Region 1: cement consumption (tons) in 2 CP wells.

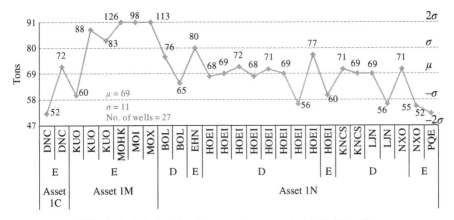

FIGURE 6.18 Region 1: cement consumption in 3 CP wells.

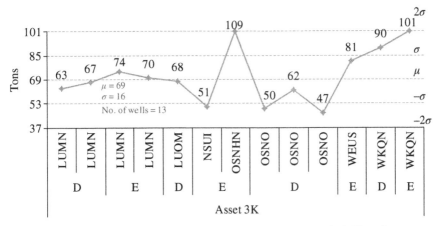

FIGURE 6.19 Region 3 (Asset 3K): cement consumption in 3 CP wells.

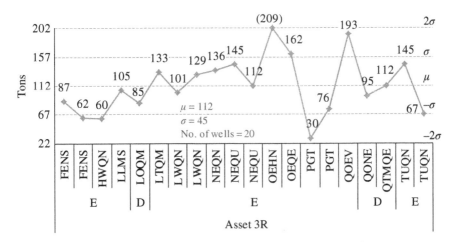

FIGURE 6.20 Region 3 (Asset 3R): cement consumption in 3 CP wells.

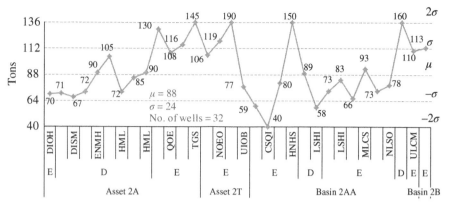

FIGURE 6.21 Region 2: cement consumption in 3 CP wells.

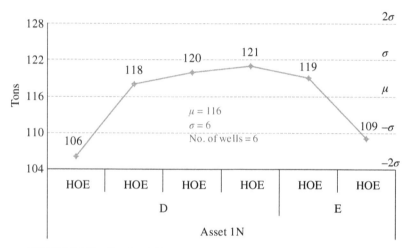

FIGURE 6.22 Region 1 (Asset 1N): cement consumption in 4 CP wells.

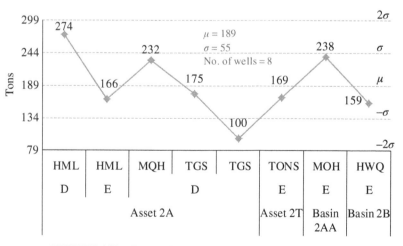

FIGURE 6.23 Region 2: cement consumption in 4 CP wells.

Offshore

(i) 3 CP wells drilled *by jack-up rigs*: $\mu = 177$ tons, $\sigma = 48$, and $N = 18$ (refer to Fig. 6.24; N, number of wells)

(j) 4 CP wells drilled *by jack-up rigs*: $\mu = 210$ tons, $\sigma = 43$, and $N = 34$ (refer to Fig. 6.25; N, number of wells)

The high standard deviation in offshore 3 CP and 4 CP wells denotes that more well data are required to fix cement consumption standard.

4 CP and 5 CP wells drilled by drillships: Sufficient well data were not available to determine the average cement consumption.

Y-axis gridlines in the following figures depict μ, $\pm\sigma$, and $\pm 2\sigma$ values and ranges. In the X-axis of these figures, D denotes development well and E exploratory well, and others are Asset/Basin names and field names (e.g., HNA, LMBB, MNCE, TES, and OETO).

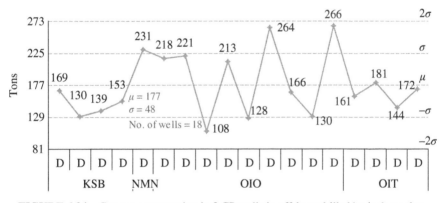

FIGURE 6.24   Cement consumption in 3 CP wells in offshore drilled by jack-up rigs.

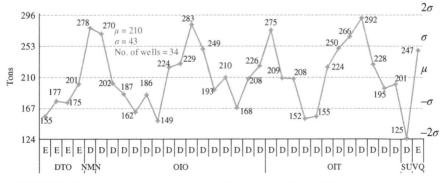

FIGURE 6.25   Cement consumption in 4 CP wells in offshore drilled by jack-up rigs.

6.5.3 Suggestions

Based on the discussion in Sections 6.5.1 and 6.5.2, the following cement consumption standards (μ) are determined for wells of similar casing policy:

(a) *Onshore*

    2 CP wells

      (i) **Region 1**: $\mu = \mathbf{40}$ tons, $\sigma = 10$

    3 CP wells

      (ii) **Region 1**: $\mu = \mathbf{69}$ tons, $\sigma = 11$

      (iii) **Region 3** (Asset 3K): $\mu = \mathbf{69}$ tons, $\sigma = 16$

      (iv) **Region 2**: $\mu = \mathbf{88}$ tons, $\sigma = 24$

    4 CP wells

      (v) **Region 1** (Asset 1N): $\mu = \mathbf{116}$ tons, $\sigma = 6$

      (vi) Cement consumption standards for 3 CP, 4 CP, and 5 CP wells for other regions/Assets could not be fixed either due to insufficient number of such type of wells in that region or nonavailability of adequate well data. 2 CP wells are drilled only in Region 1.

(b) *Offshore*

      (vii) The wide variations and high standard deviation in offshore 3 CP and 4 CP wells drilled by jack-up rigs denote that more well data are required for standardizing cement consumption. Likewise, sufficient well data were not available to determine the average cement consumption for 4 CP and 5 CP wells drilled *by drillships*.

(c) Allowances may be provided for about 3–5% wells drilled in a year, which consume high quantity of cement under exceptional circumstances such as additional cement jobs due to well complications, mud loss, repair jobs, sidetracking, and so on. Higher allowances may be considered, if the procurement lead time of oil well cement is longer.

(d) The standard consumption of cement (as well as HSD and chemicals) so determined is preliminary only. This may be reviewed annually incorporating consumption data of additional wells, so that with the increase in sample size, standards become more reliable.

6.6   STANDARDIZATION OF CHEMICAL CONSUMPTION

The basic ingredients of drilling fluids are chemicals, which are expensive. This study attempts to standardize consumption of chemicals in oil and gas well. But before proceeding further, it would be useful to know the composition, properties, and functions of drilling fluids, which are elaborated in the following paragraphs.

Drilling fluids, generally known as "mud," are considered as "lifeline" of drilling operation. They are basically slurry of chemicals, additives, and water, which is pumped in at a predetermined rate and pressure through the drill pipe to the bottom of the hole. It comes out from the nozzles of the drill bit through the annulus of the wellbore and drill pipe and reaches to the surface. In the process, drilling fluids performs several important functions that are mentioned in the following paragraphs [1]:

(i) Removes rock cuttings from the bottom of the hole and transports them to the surface, thus ensuring further drilling.

(ii) Controls formation pressure—as hydrostatic pressure of mud column balances subsurface pressure.

(iii) Cools and lubricates drill bit and drill string, as considerable heat is generated due to friction at the drill bit, and between drill string and wellbore.

(iv) Ensures suspension of solid particles, which is helpful, especially during tripping, making connections and downtime.

(v) Seals off permeable formations and stabilizes wellbore by forming filter cake on the borehole wall to minimize fluid invasion and formation damage.

(vi) Controls corrosion.

(vii) Transmits hydraulic horsepower to the bit; hydraulic horsepower to the bottom of the well may be maximized by choosing appropriate bit nozzle size.

(viii) Facilitates collection of formation data—as important geological information and drilling parameters are obtained through mud log, apart from facilitating MWD and LWD.

(ix) Partially supports the weight of drill string and casing, as buoyancy equal to the weight of displaced drilling fluid offsets the weight of drill string and casing.

(x) Assists in cementing and completion jobs.

Drilling fluids are broadly classified into three categories, namely, (i) *water-based mud*, (ii) *oil-based mud*, and (iii) *aerated drilling fluids*. **Water-based mud** is the most commonly used mud in which clays, chemicals, and additives are added into water to prepare a homogeneous blend with desired properties for shale stability, viscosity control, cooling and lubricating of equipment, increased ROP, and so on. The main components of water-based mud are bentonite with additives such as Barites (barium sulfate), calcium carbonate, polymers, shale stabilizers, and other chemicals for obtaining or controlling desired mud properties.

In **oil-based mud**, the base fluid is generally a petroleum product, for example, diesel oil, and it is used for higher lubricity, increased shale stability, low formation pressure, and faster drilling. It has greater resistance to temperature in HT wells, less viscosity, lesser friction, and better steerability for directional hole. It is reusable and helps in corrosion control, but it is environmentally sensitive and costly.

Aerated drilling is used under special circumstances such as underbalanced/managed pressure drilling to prevent formation damage and lost circulation and to increase penetration rate in hard-rock areas. It has many variants, namely, dry air drilling, mist or foam drilling, and aerated mud drilling, which are used under different conditions and circumstances, while the operating principles are the same.

The selection of drilling fluid depends on several factors such as (i) formation pressure and temperature, (ii) lithology, (iii) availability of soft water, (iv) possibility of encountering connate water, (v) economics, and so forth.

6.6.1 Fixing Standards for Consumption of Chemicals

There are about 40 chemicals used for drilling fluids. Different set of chemicals are used for different mud policies, depending on field properties, structures, formation pressure, and so on. ABC analysis of chemicals used for drilling fluids in offshore reveals that Barites account for nearly 30% of chemicals cost, followed by XC polymer (12%), KCL (10%), and remaining 35 chemicals (28%). This is graphically shown in Figure 6.26. Similarly, in onshore the cost of consumption of Barites accounts for maximum, followed by bentonite, caustic soda, KCl, and so on. Therefore, standardization has been limited to the consumption of Barites, XC polymer, and CMC in offshore and Barites, bentonite, and caustic soda in onshore. The cost of chemicals accounts for 1.8% of the well cost in offshore.

Consumption of chemicals in an oil/gas well depends on the following:

(i) Depth of well/meterage or footage drilled
(ii) Formation pressure
(iii) Lithology of the well
(iv) Borehole instability

It is often found that the actual values of some of these factors vary extensively from the estimate, causing standardization of chemical consumption difficult. The unit consumption, that is, **consumption of chemicals per meter drilling**, has been chosen as parameter for fixing standards of similar type of wells in a field/Asset/Basin based on historical data. However, wells incurring mud losses and consuming excessive chemicals have been excluded for the purpose of standardization.

6.6.2 Observations

Chemical consumption data of about 332 wells drilled during 1–1½ years in onshore and offshore were analyzed. The consumption of chemicals (Barites, XC polymer, bentonite, and caustic soda) per meter drilling was determined. The analysis of unit chemical consumption covering different wells/fields/regions reveals the following:

(a) *Except for a few fields in Asset 3K and other Assets, there is a wide variation in unit consumption (kg/m) of chemicals in wells within the same field with high standard deviation.* This is because of wellbore instability that requires unplanned mud weight increase and higher doses of costly shale-stability additives.

(b) The variations are alike for both exploratory and development wells within the same field, Asset, or Basin. *Therefore, instead of determining standard unit consumption of chemicals, ranges of consumption of a group of wells with an upper limit of consumption have been specified.*

(c) The average consumption of Barites per meter drilling in different fields of Asset 3K was found as $\mu = 35$ kg/m, with $\sigma = 18$.

(d) The consumption of chemicals is largely dictated by well types and shale types. The consumption pattern of other chemicals (XC polymer and bentonite) is found to be similar to that of Barites.

(e) The unit consumption of Barites in different fields/Assets of on-land Regions 1, 2, and 3 and offshore are shown in Figures 6.27, 6.28, 6.29, and 6.30, respectively.

In X-axis of these figures, D denotes development well and E exploratory well, and the others are Asset/Basin names and field names (e.g., BNE, HNA, LMBB, MNCE, and HOEI).

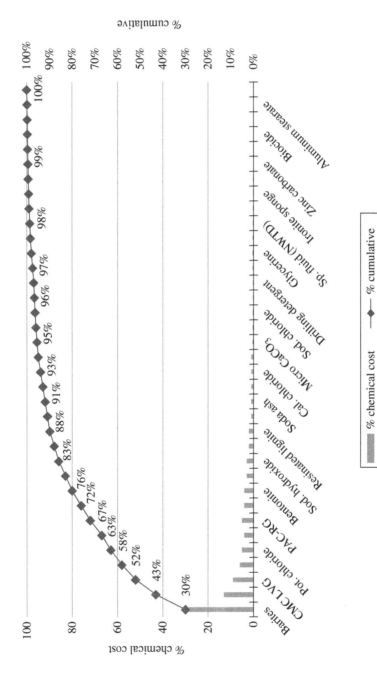

FIGURE 6.26 ABC analysis of chemicals used in offshore.

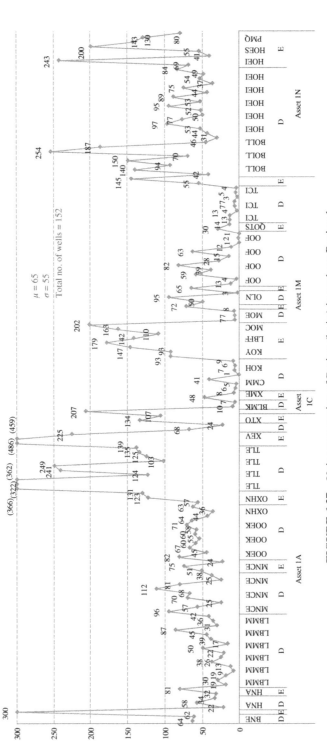

FIGURE 6.27 Unit consumption of Barites (kg/m) in onshore Region 1.

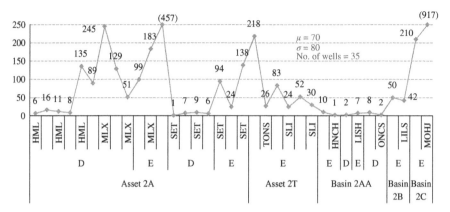

FIGURE 6.28 Unit consumption of Barites (kg/m) in onshore Region 2.

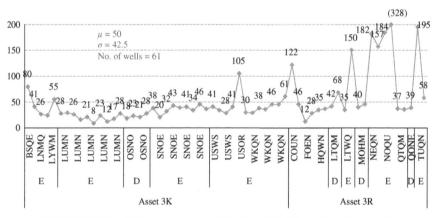

FIGURE 6.29 Unit consumption of Barites (kg/m) in onshore Region 3.

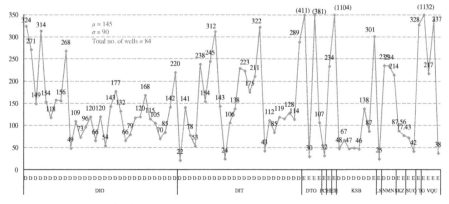

FIGURE 6.30 Unit consumption of Barites (kg/m) in offshore.

6.6.3 Suggestions

Based on the observations and analysis in Section 6.6.2, the following suggestions are made, which would help in specifying range and upper limit of unit chemical consumption in various fields/Assets/regions:

(a) Except for few fields, there is a wide variation in unit consumption of chemicals even in similar type of wells within the same field/Asset/region. Therefore, instead of fixing standards for chemical consumption in these fields/Assets/regions, ranges and upper limit of consumption in wells have been determined.

(b) For the purpose of planning, monitoring, and procurement in different Assets/ Basins/regions, the following can be considered:

    (i) 76% of wells in onshore Region 3 consume Barites below (<) 50 kg/m with an average consumption of $\mu=31.6$ kg/m, followed by 11% of wells consuming 50–100 kg/m with $\mu=70$ kg/m, and the remaining 13% of wells consume more than 100 kg/m (Fig. 6.31).

    (ii) The same data for other regions are summarized in Table 6.1 and depicted in Figures 6.32, 6.33, and 6.34.

    (iii) Not more than 10% wells in onshore Region 3, 11% wells in onshore Region 1, 17% wells in onshore Region 2, and 39% wells in offshore consume Barites more than 150 kg/m.

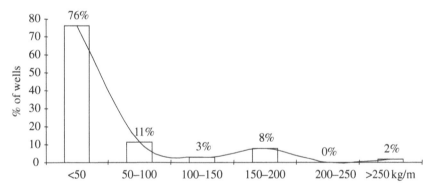

FIGURE 6.31   Pattern of Barite consumption in onshore Region 3.

TABLE 6.1   Percentage of Wells Consuming Barites within Ranges

Region	<50 (kg/m)	50–100 (kg/m)	100–150 (kg/m)	>150 (kg/m)
Onshore Region 3	76%	11%	3%	10%
	$\mu=32$	$\mu=70$	$\mu=126$	
Onshore Region 2	60%	15%	8%	17%
	$\mu=15$	$\mu=78$	$\mu=134$	
Onshore Region 1	44%	33%	12%	11%
	$\mu=23$	$\mu=70$	$\mu=130$	
Offshore	16%	17%	28%	39%
	$\mu=38$	$\mu=74$	$\mu=124$	

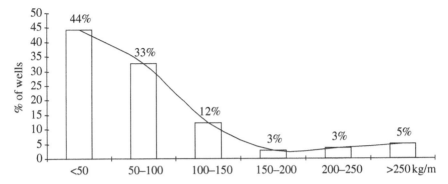

FIGURE 6.32   Pattern of Barite consumption in onshore Region 1.

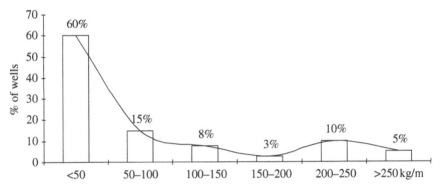

FIGURE 6.33   Pattern of Barite consumption in onshore Region 2.

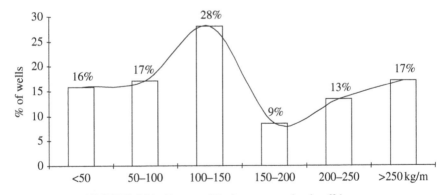

FIGURE 6.34   Pattern of Barite consumption in offshore.

6.7   BENEFITS AND SCOPE FOR IMPROVEMENT

There is scope of **savings up to 62.8 million** *on account of fuel consumption and enhancing life of engines.* Apart from this, fixing standards for consumption of HSD, cement, and chemicals will help in (i) preventing stock-out situation, (ii) avoiding building excess inventory, (iii) monitoring consumption, (iv) checking wastage and aberrations, and (iv) drawing plan for procurement.

6.7.1 Scope for Improvement

The consumption standards fixed for HSD, cement, and chemicals in this study may be considered as initial standard value, which can be further refined with more consumption/well data. This will enhance the credibility and robustness of the standards developed in this study.

The data required for standardizing consumption of HSD for different capacity of drilling rigs in different geologic conditions/fields/regions are not sufficient to prove statistical robustness. The reliability of standards would increase if sample size is enlarged, that is, more consumption and well data are available. Since more wells have been drilled in these fields using similar capacity rigs in subsequent years, the additional consumption/well data can be incorporated to make the standard statistically significant. Similarly, cement consumption standards for 2 CP, 3 CP, 4CP, and 5 CP wells at different fields/regions can be updated with additional well data. Likewise, chemical consumption standard can be reevaluated with additional data for statistical significance and robustness. Further work can be done in this regard to arrive at reliable and conclusive standards.

6.7.1.1 Workover Operations: Workover jobs are of varied types and are different from drilling operation. Further work may be undertaken to standardize fuel consumption of workover rigs/jobs in the future, as it varies from that of drilling rigs.

6.8 CONCLUSIONS

The study aims to standardize consumption of HSD, cement, and chemicals in oil/gas wells and rigs. The findings and suggestions are summarized in the following text.

6.8.1 Standardizing HSD Consumption

The fuel consumption in drilling rigs depends on various factors such as rig type and capacity, nature of operations, lithology, ROP and depth, fuel efficiency of engines and equipment, and so on. Rigs of similar types and capacity engaged in similar operations in similar geological formation and depth were grouped together, and "consumption per meter drilling" has been adopted as parameter for fixing standard, as explained in Section 6.4.1.

Accordingly, average unit fuel consumption (μ) for each type of drilling rigs in different fields/Assets/regions was computed using consumption data for last 2–3 years. Fuel consumption standards for various capacity rigs operating in Region 1 and 3 are found to be consistent with low standard deviation. But fuel consumption of similar rigs in Region 2 is found to be much higher than that in Region 1 and 3 with high dispersion (standard deviation). The reasons for such wide variations have been investigated, and finally, suggestions have been offered to minimize wastage and improve unit fuel consumption.

Energy wastage in drilling rigs can be substantially reduced by the following:

(i) Efficient use of engines, mud pumps, agitators, air compressors, and so on, by following standard operating practices.

(ii) Improving load on engines from abysmally low (~20–45%) to optimum load (70–90%) on engines. HSD consumption will considerably reduce if the load on engine is improved to optimal or desired level. Besides, it would enhance life of at least one engine per drilling rig.

(iii) Installing load monitoring system and improving maintenance of engines and equipment.

(iv) Replacing energy-inefficient engines with new-age engines.

(v) Appropriate deployment of rigs: Avoid deploying higher-capacity rigs for shallow wells that cause significant variations in fuel consumption. Exceptional circumstances such as prolonged downhole complications, production testing, and so on (leading to low meterage/footage drilled), are to be excluded while fixing fuel consumption standard.

HSD consumption of drillships and some category of on-land rigs could not be standardized due to (i) wide variation in capacity and applications and (ii) inadequate number of such category rigs and inadequate data.

6.8.2 Standardization of Oil Well Cement Consumption

The consumption of cement in a well depends on various technical factors such as hole size and depth, casing policy (i.e., number and sizes of casing), cement rise, number of prospective objects to be tested, and downhole complications (i.e., activity/mud loss, caving, sidetracking, etc). It was observed that the consumption of cement in *wells of similar casing policy, hole size, and depth in the same field/region varies within an acceptable range with low standard deviation, irrespective of the type of wells*. Therefore, the consumption of cement for different casing policy wells in the same field/region was standardized for both onshore and offshore wells.

6.8.3 Standardization of Chemical Consumption

The consumption of chemicals in a oil/gas well depends on various geological and technical factors such as well depth/meterage or footage drilled, formation pressure, lithology and subsurface properties, borehole instability, and so on. The consumption of "chemicals per meter drilling" has been chosen as parameter for fixing standards, as explained in Section 6.6.1. It was noticed that except for a few fields in one of the Assets, there is a wide variation in unit consumption (kg/m) of chemicals in wells within the same field with high standard deviation. Therefore, instead of determining standard unit consumption of chemicals, consumption range of a group of wells with upper limit of consumption has been determined.

6.8.4 Suggestions and Benefits

The standard consumption of HSD, cement, and chemicals, as determined in this study, may be considered as a preliminary value for further study. These may be reviewed periodically incorporating consumption data of additional wells drilled in the intervening period, so that with the increase in sample size, standards become more reliable and robust. Finally, suggestions have been offered for improvement in consumptions of HSD, cement, and chemicals, which would help in forecasting and monitoring consumption, planning procurement and preventing stock-out, avoiding excess inventory, and minimizing wastage. The detailed suggestions are listed in Sections 6.4.4, 6.4.6, 6.5.3, and 6.6.3. These suggestions have the potential of savings up to USD 62.8 million on account of fuel consumption and enhancing life of engines.

Chapter 7 deals with optimization of rig move time and develops optimal activity schedule using critical path analysis. Since rig idling cost is very high, there is continuous endeavor to minimize unproductive rig time. Thus, optimizing rig move time would ensure more productive time for drilling, which would result in higher productivity of drilling operation.

REVIEW EXERCISES

6.1 Why it is necessary to standardize the consumption of diesel in drilling rigs? What are the factors that influence the consumption of fuel in drilling rigs?

6.2 Why does fuel consumption vary in different regions/fields for the same capacity rigs?

6.3 How would you improve fuel efficiency of (diesel) engines at drill sites? How would you standardize diesel consumption in drilling rigs? Why it is difficult to standardize fuel consumption in workover rigs?

6.4 Why it is necessary to standardize cement consumption in oil and gas wells? What are the factors considered for determining cement consumption in a well?

6.5 How would you standardize consumption of cement in oil and gas wells?

6.6 What are the functions and techniques of primary cementing?

6.7 What are the purpose and techniques of secondary cementing?

6.8 What are the classifications of oil well cement?

6.9 What are the benefits of standardization of consumption of chemicals in oil and gas wells? What are the factors that control the consumption of chemicals in a well?

6.10 How would you standardize the consumption of chemicals in oil and gas wells?

6.11 What are the main functions of drilling fluids?

6.12 What are the various types, composition, and properties of drilling fluids?

Advance

6.13 As a cementing specialist, what are the factors you would consider in designing cement slurry?

6.14 What are the selection criteria for drilling fluids? As a mud engineer, how would you design drilling fluids for a well (consider different types of well)?

6.15 What is circulation loss? As a drilling/mud specialist, what are the steps you would take to control it?

APPENDIX 6.A   DETAILS OF OWN AND HIRED RIGS DEPLOYED AT OFFSHORE ASSET

Rig Name	Rated Water Depth (ft)	Drilling Depth (ft)	Draw Work Capacity (HP)	Derrick Capacity (klb)
Own jack-up rigs				
O/Jack-up 1	250	18,000	2000	1100
O/Jack-up 2	300	20,000	2000	1392
O/Jack-up 3	350	25,000	2000	1392
O/Jack-up 4	300	20,000	2000	1392
O/Jack-up 5	300	20,000	2000	1392
O/Jack-up 6	300	20,000	2000	1044
O/Jack-up 7	300	20,000	2000	1392
O/Jack-up 8	300	20,000	2000	1392
Hired jack-up rigs				
H/Jack-up 1	300	25,000	2500	1392
H/Jack-up 2	300	20,000	2000	1000
H/Jack-up 3	300	25,000	3000	1044
H/Jack-up 4	300	25,000	3000	1000
H/Jack-up 5	300	20,000	1900	1350
H/Jack-up 6	300	25,000	3000	1300
H/Jack-up 7	300	20,000	2000	1000
H/Jack-up 8	300	20,000	2000	1400
H/Jack-up 9	300	25,000	3000	1000
H/Jack-up 10	300	21,000	2000	1400
H/Jack-up 11	300	25,000	2000	1300
Own drillship				
O/Drillship 1	1000	20,000	2000	1400
O/Drillship 2	2950	20,000	3000	1400
Hired drillship				
H/Drillship 1	1150	25,000	2400	1000
H/Drillship 2	2000	20,000	2000	825

REFERENCE

Useful Link

[1] Nguyen, T., Drilling Engineering, Introduction to Drilling Fluids, available at: https://www.google.com/?gws_rd=ssl#q=Drilling+Engineering+%E2%80%93+PE+311++Chapter+2:+Drilling+Fluids (accessed on February 29, 2016).

FURTHER READING

Bommer, P., A Primer Oil Well Drilling: A Basic Text of Oil and Gas Drilling, 7th edition, The University of Texas, Austin, TX, 2008, in cooperation with International Association of Drilling Contractors (IADC).

Bourgoyne, A.T., Jr., Millheim, K.K., Chenevert, M.E., and Young, F.S., Jr., Applied Drilling Engineering, SPE Text Book Series, Vol. 2, Society of Petroleum Engineers, Richardson, TX, 1986.

Chilingarian, G.V. and Vorabutr, P., Drilling and Drilling Fluids, Elsevier Science Publisher, Amsterdam, 1983.

Gray, G.R. and Darley, H.C.H., Composition and Properties of Oil Well Drilling Fluids, 4th edition, Gulf Publishing Co., Houston, TX, 1980.

Nelson, E.B. and Guillot, D., Well Cementing, 2nd edition, Schlumberger, Houston, TX, 2006.

Nind, T.E.W., Principles of Oil Well Production, 2nd edition, McGraw-Hill Book Co. Ltd., New York, 1981.

Serada, N.G. and Solovyov, E.M., Drilling of Oil and Gas Wells, MIR Publisher, Moscow, 1977.

Smith, D.K., Cementing, Society of Petroleum Engineers, Richardson, TX, 1989.

Useful Link

IPIECA, Offshore Drilling Rigs, Energy Efficiency Practices Compendium, 2013, available at: http://www.ipieca.org/energyefficiency/solutions/60311 (accessed on March 20, 2016).

7

OPTIMIZING RIG MOVE TIME AND ACTIVITY SCHEDULE USING CRITICAL PATH ANALYSIS

7.1 INTRODUCTION

Earlier in this book, we have dealt with various aspects of drilling operation including components of productive and nonproductive drilling time and suggested measures to optimize controllable rig time loss and drilling productivity. In this chapter, we would focus on optimizing rig move time, which is part of rig cycle time.

Deep drilling is the most expensive E&P activity that requires specialized manpower and technology. It is the constant endeavor of E&P companies and drilling operators to contain spiraling drilling cost and avoid idling of rigs by various means. They are hard pressed to minimize nonproductive rig time and improve rig time availability for drilling. In this context, the current study focuses on optimization of rig move/mobilization time (which is considered as unproductive rig time), so as to ensure more time for on-bottom drilling and completion.

After completion of a well, on-land rig moves to the next location for drilling. *Rig move/ mobilization is part of rig cycle time and consists of rig-down, transportation, and rig-up activities.* The rig is dismantled after completion of a well, which is called "rig-down"; rig equipment are moved to the next location called "relocation" or "transportation," where it is assembled again called "rig-up" for commencing drilling operation at the new location/ well. Rig cycle is the time between rig release from the current well (location) and rig release of the next well (location). *It has three main components, namely, drilling, production testing, and rig move/mobilization activities.* Rig move/mobilization is the time between rig releases from the just completed well and spud of the new well, and is considered as unproductive rig time.

Optimization and Business Improvement Studies in Upstream Oil and Gas Industry, First Edition. Sanjib Chowdhury.
© 2016 John Wiley & Sons, Inc. Published 2016 by John Wiley & Sons, Inc.

Before we delve deep into the current study for optimization of rig move/mobilization time and activity schedule, it would be useful to know various types of on-land rigs and their utility, which are described in the following paragraphs.

7.1.1   A Brief Note on Drilling Rigs

Drilling operation is mainly of two types, namely, "exploratory" and "development," depending on the purpose and type of wells. Exploratory drilling is usually carried out in the virgin area or in the field whose lithology or subsurface characteristics are unknown. On the other hand, development drilling is undertaken in a producing field with known reservoir characteristics to enhance production capacity. Based on capacity, depth of drilling, and size of production hole, rigs are categorized as "shallow" or "deep drilling" rig.

Similarly, depending on the type of operation performed, on-land rigs are termed as "drilling" or "workover" rigs. Old wells occasionally require maintenance and recompletion jobs, especially in matured fields that include revival of sick wells, water shutoff jobs, well deepening, sidetracking, fishing, servicing, and so on, that necessitate working over the wells. These jobs are generally done with smaller-sized, cheaper rig with lesser footprint called "workover rig."

A deep drilling rig consists of large-sized modules and high-capacity equipment like draw work, engine, mud pump, mud system, traveling block, crown block, swivel, rotary table, and huge and tall tower called "derrick," apart from varied downhole tubular such as drill pipe, collar, stabilizer, jar, and so on. A shallow or development drilling rig has a smaller-sized module as compared to a deep drilling rig. Workover rigs, on the other hand, are generally truck mounted and have much lesser equipment.

The on-land drilling rigs are of various types, which may be broadly termed as mobile rig, desert rig on wheels, and rig with high floor mast and substructure, all of which are elaborated as follows:

1. **Mobile rig**: This type of rig is mounted on wheeled carrier and has telescopic mast, which can be driven to the desired location along with equipment, engines, and other accessories. It is used for drilling shallow wells. Rig move time ranges between 16 and 24 h within a distance of 4–5 km.
2. **Desert rig on wheels**: There are various combinations of rig types depending on capacity, need, varying (heavy) loads on wheels, partially or fully disassembled units, and so forth for which rig move time would also vary. Desert rig on wheels that moves with mast in erect condition takes around 1½–2 days for rig move within the same area. Desert rig that moves with mast down condition takes around 3½–4 days.
3. **Rigs with high floor mast and substructure**: These are usually higher-capacity rigs used for drilling deeper wells. The rig components are dismantled and transported to the new location by heavy-duty trucks and trailers. Rig move time varies from 1 to 2 weeks depending on the design and capacity of the rig. The current study specifies rig move activities of 1500/2000 HP desert electrical rig on wheels *with fully disassembled units* (drilling depth capacity—16,000/20,000 ft).

7.2   OBJECTIVES

Studies carried out by various agencies reveal that rig move/mobilization accounts for a substantial amount of average rig activity time. As per a study conducted by Alvarez and Marsal in 2009, rig mobilization time consumes around 25% of average rig activity time [1, 2]. For this purpose, there is a need to standardize rig move time for various categories of rig and optimize rig move activity schedule. Toward this end, the objectives of this study are as follows:

(a) To develop optimal schedule of rig move/mobilization activity for on-land deep drilling rigs
(b) To optimize and standardize rig move/mobilization time that may be used as norm time for preparation of drilling plan and incentive purpose
(c) To suggest measures for improving efficiency of rig mobilization and use critical path analysis for decision making for resource allocation, "what-if" tests, crashing activities, alternative schedule, and so on

Rig move time also depends on the design of rig, but this aspect is beyond the scope of this study. In fact, rig manufacturers are engaged in continuous improvement of rig design with emphasis on safety, operational performance, and employee productivity.

7.2.1   Methodology

This study uses network analysis based on Program Evaluation and Review Technique and Critical Path Method (PERT/CPM) technique to meet the aforementioned objectives for on-land deep drilling rigs of higher capacity. The following approach has been adopted to develop the network model:

(i) Identifying and listing rig move/mobilization activities (i.e., rig-down, relocation, and rig-up)
(ii) Defining relationship and sequence of activities including predecessors and followers
(iii) Estimating duration or completion time of each task/activity
(iv) Assigning resources (i.e., manpower, crane, trailer, etc.) required for each task
(v) Drawing activity network, computing critical path, and identifying critical activities

The network analysis is commonly known as critical path analysis and is widely used in project management. The detailed discussion on critical path analysis and the use of PERT/CPM technique are discussed in the following paragraphs.

7.3   A BRIEF NOTE ON NETWORK ANALYSIS USING PERT/CPM

PERT and CPM are essentially network analysis techniques for planning, scheduling, monitoring, and controlling of project tasks/activities, so that the project is completed within the specified time and cost [3–5]. These techniques are used in complex project involving a large number of activities with interdependencies and are popular project management tools. PERT and CPM were developed nearly at the same time by two independent groups in the United States for different purposes. CPM was developed by DuPont Corporation in association with RAND in 1956/1957 for construction of a new chemical plant and preparation of maintenance shutdown schedule. J. E. Kelly of RAND and M. R. Walker of DuPont are said to be the main architects for developing the CPM. *The emphasis of the approach was on balancing the project cost and the project completion time.*

PERT was developed in 1957/1958 by the US Navy in consultation with the consultancy firm Booz Allen Hamilton for managing the plan for POLARIS missile program. It was a large, complex engineering and development project involving around 250 major contractors and 9000 subcontractors. *The emphasis was on completing the program in the shortest time possible.* Both these techniques gained prominence with the success of these projects. Later on, they were used in diverse projects across industry and other sectors, such as construction of factories, buildings, bridges, dams, roads, ports, rails, ships, and airports; space mission; research and development; and so on.

Although these methods (PERT/CPM) were developed independently, in due course of time differences narrowed down. The practicing firms adopted the best features of both these techniques to plan and manage their projects, and with the passage of time, they became synonymous. But still, there are distinct differences, which are as follows: *CPM is an activity-based* network and the *task times are deterministic* (one time estimate), that is, it can be estimated with a fair degree of accuracy. It is mostly used in construction projects and product management where jobs are of *repetitive nature* and the emphasis is on *managing completion time and cost.* CPM uses *activity-on-node* network construction, wherein nodes denote activities and arrows depict precedence relationship.

On the other hand, *PERT is an event-based* network and is mostly used in *nonrepetitive nature* of jobs such as research and development projects. The *task times are probabilistic with three time estimates,* namely, optimistic, most likely, and pessimistic time for arriving at the expected completion time. The emphasis is on completion time, since cost is assumed as directly proportional to time. PERT uses *activity-on-arrow* network construction, wherein arrows denote activities and nodes symbolize events at a point in time where single or multiple activities can start and/or finish.

A project consists of thousands or hundreds of tasks/activities that are carried out in sequence and/or in parallel. It is not possible to focus equally on all activities, as they have varying degree of importance, completion time, resource requirement, and so on. Although a large project may have thousands of activities, only few (say, hundred) can be termed "critical," which have bearing on the project completion time. These activities need to be identified and monitored closely for timely completion of the project. Any delay in critical activities would extend the project completion time, resulting in an escalation of project cost. Therefore, focus should be to control the critical activities that would ensure successful completion of the project.

The critical path has least flexibility, and if any critical activity is delayed, it would delay the project completion time by the same amount. On the other hand, noncritical activities have some flexibility; delay in noncritical activities will have no impact on the project completion time, if such delay is less than the slack. In case delay in noncritical activities exceeds the slack, the project will be delayed to the extent of exceeding the slack time.

Critical path analysis using PERT/CPM technique is a popular tool for project management and is widely used nowadays for decision making, finding alternatives, and optimizing project completion time and cost. As we proceed further, we would come across various elements of critical path analysis and steps for developing network analysis model. One would also get familiarized with forward pass, backward pass, earliest start time, earliest finish time, latest start time, latest finish time, slack, float, critical path, critical activities, dummy activities, node/events, normal time, crash time, crash cost, probability of completion of project within the due date, and so on.

7.4 DEVELOPMENT OF NETWORK ANALYSIS MODEL

A network is characterized by a sequence of activities, interdependencies, interrelationship, and criticality of activities. It is a graphical representation of a set of activities (called project) illustrating the flow and sequence of constituent activities and events. There are two types of network construction/diagram, namely, (i) **activity-on-node**, where nodes denote activities and arrows depict precedence relationship (used by CPM), and (ii) **activity-on-arrow**, where arrows denote activities and nodes symbolize events at a point in time (used by PERT). Arrows point in the direction of progress of the project.

The network analysis is a powerful approach for minimizing project completion time, cost, and idle resources. *The network model is essentially a set of events or activities that consume time, labor, and physical resources.* It is the application of organized planning, scheduling, and controlling of several events or activities that take place sequentially and/ or simultaneously.

Scheduling of Rig Move Activities Using PERT/CPM: PERT/CPM technique has been used in this chapter to develop optimal schedule of rig move activity and standardize rig move time of on-land 1500/2000 HP desert electrical rig on wheels with *fully disassembled units*. This was done after studying several rig move events, consulting with experts, and validating tasks, relationship, and interdependency by rig move professionals. All these require thorough knowledge and expertise of rig move work. The steps mentioned in Section 7.2.1 were followed, which are elaborated in the following text.

7.4.1 Identifying Tasks/Activities

The first step is to identify the tasks/activities for rig-down, transportation, and rig-up. There are hundreds of activities during rig mobilization, but for the sake of convenience, better control, and ease in implementation, several small tasks have been clubbed as one task. The consolidated list of tasks during rig move/mobilization (rig-down and rig-up) is shown in Appendix 7.A.

A deep drilling rig consists of various modules, equipment, and machineries, which can be broadly categorized into three subsystems, namely, substructure, mast, and other accessories. Accordingly, the activities during rig-down and rig-up have been divided into three subgroups for ease of understanding, control, and modeling. Appendix 7.A also contains the activity descriptions.

7.4.2   Defining Relationship and Interdependencies of Tasks/Activities

The next step is to find the relationship among various tasks/activities, their interdependencies, sequence of events, and predecessors and followers of each event. This is an important step, and due consideration may be given to define the relationship and sequence of events. The predecessor(s) of each activity for rig-down and rig-up is shown in Appendix 7.A from which interdependencies and sequence of events may be derived. Sometimes, the task relationship may need further refinement.

7.4.3   Estimating Duration of Tasks/Activities

The completion time of each task is required to be determined, which is usually done by work study, benchmarking, analysis of historical data, expert's opinion, experience, and so forth. Three time estimates, namely, optimistic (T_o), most likely (T_m), and pessimistic (T_p) times, are considered for arriving at expected completion (T_e) time of each task in PERT, which essentially follows β-distribution and is denoted as

$$T_e = \frac{T_o + 4T_m + T_p}{6} \qquad (7.1)$$

$$\text{Variance of each activity } (\sigma^2) = \left\{ \frac{(T_p - T_o)}{6} \right\}^2 \qquad (7.2)$$

and square root of variance is the standard deviation (σ) of expected completion time of each activity.

The duration of each task/activity during rig move is shown in Appendix 7.A, which is essentially the expected completion time (T_e). It is important that the planned completion time of each task/activity is realistic and is not too loose or tight, as this has direct bearing on the project completion time (i.e., rig move time). The activity completion time also depends on resources (i.e., manpower) assigned to each task/activity, which is elaborated in Section 7.4.4.

7.4.4   Assigning Resources for Tasks/Activities

It is important to define the resource requirement for each task/activity, as it has direct influence on completion time and cost of the activity/project. For example, if a task with two persons takes 2 h to complete, it would likely to take 4 h to complete with one person or 1 h with four persons (considering linear functionality). But, in most of the cases, resources are limited, be it manpower, machinery, capital, or others. Therefore, it is essential to assign scarce resources prudently to get maximum benefit. The algorithm for resource allocation tries to assign limited resources in the best possible manner among competing tasks/activities, so that the project is completed with minimum delay or in minimum possible time. *The criteria, procedure, and steps that are generally followed for allocation of limited resources among competing tasks/activities have been described in detail in Section 7.9.1.*

Usually, a fixed amount of money is paid for rig mobilization; in addition, incentive is paid for completing rig move before the norm time. *Because of economic reasons, E&P*

companies and drilling contractors arrange adequate resources for rig move activities, so that delay in rig mobilization is minimized. The main resources for rig move activities are rig crew (manpower), cranes (heavy equipment for material handling), and trailers (transport fleet). It has been observed in this study that in-house rig crew are deployed in adequate numbers and sufficient number of cranes, trailers, and transport fleet are arranged for rig move/mobilization activity.

The current study makes a pragmatic assumption that there are no constraints for resources, that is, sufficient rig crew, cranes, and trailers are available. The resource requirement (manpower) for each task for rig move activity is shown in Appendix 7.A.

7.4.5   Drawing Activity Network and Determining Critical Path

After identifying all tasks/activities of rig move, defining interdependencies and relationship among tasks, estimating duration of each task/activity, and assigning resources for each task, we are ready to draw the activity network and determine the critical path of rig move operation, which are elaborated in the following text.

7.4.5.1   Computing Critical Path: The computation of critical path involves forward pass as well as backward pass through the network. The **forward pass** is the process of moving along a network computing the earliest start (ES) time and earliest finish (EF) time for each activity from beginning to end. In other words, it deals with the ES and EF time of all tasks and moves from start to finish event.

The **backward pass** is the process of moving backward along a network computing the latest finish (LF) time and latest start (LS) time for each activity from end to beginning. To put it simply, it deals with the LS and LF time of all tasks and proceeds backward from finish to start event. The steps for computation of forward pass and backward pass are explained as follows:

Forward Pass
(a) In order to make a forward pass through the network, identify all activities that have no predecessors:
   - The earliest start time of all such activities is zero, that is, $ES = 0$.
   - The earliest finish is basically the duration of activity, that is, EF = activity duration (T_e).
(b) Compute the ES of all events for which EF of immediate predecessors has been determined:
   - ES = maximum EF of immediate predecessors
   - $EF = ES + T_e$
(c) Repeat the process until all nodes/events have been computed.
   EF of the finish node/event is the earliest finish time of the project.

Backward Pass
(d) In order to make a backward pass through the network, identify all penultimate activities of finish node/event:
   - The latest finish time of such activities is LF = minimum project time.
   - The latest start of such activities is $LS = LF - T_e$.

(e) Compute LF of all nodes/events for which LS of all immediate successors have been determined:
- LF = minimum LS of immediate successors
- $LS = LF - T_e$

(f) Repeat the process backward until all nodes/events have been computed.

7.4.5.2 Computing Slack time: After calculating the ES, EF, LS, and LF, we are now in a position to compute another important variable called "slack" or "float" (see Appendix 7.B). *Slack or float is the difference between the latest allowable time and earliest completion time of an event. It is the amount of time an activity can be deferred without delaying the project completion time.*

There are mainly two types of float, namely, total float and free float. Total float is the difference between the LS time and ES time or the LF time and EF time of an activity, that is, **total float = LS − ES = LF − EF.**

Free float is the spare time available when all preceding and succeeding activities occur at the earliest possible times. When an activity has zero total float, free float will also be zero.

7.4.5.3 Identifying Critical Path: *The critical path is the sequence of activities/ events where there is no slack, that is, activities with zero slack connecting the start event with the finish event. It is the longest path through the network in terms of time and is the minimal project completion time. Activities with zero slack or float are called critical activities and lie on the critical path.* There may be more than one critical path in a network.

As explained in Section 7.4.5.2, ES, EF, LS, LF, and float have been computed for each activity and are shown in Appendix 7.B.

7.4.6 Assumptions and Prerequisites

The following assumptions have been made for the current study:

(a) The availability of cranes would not be a constraint during rig move activity. The required type and capacity cranes would be available at both rig-down and rig-up locations.

(b) There are a sufficient number of trailers and transport fleet available for the transportation of rig equipment from the old location to the new location.

(c) Rig move activities are carried out round the clock (24 h operation).

(d) Adequate number of rig crew is available for rig move activity. Additional manpower, if required, may be arranged in-house by mobilizing employees on special duty or paying overtime. In case rig move activity is restricted to daytime, then day and night shift crews are rearranged, so that there are sufficient rig crews for rig move activity.

(e) In-house resources have been considered for the entire rig move activity.

The validity of these assumptions has been explained in Section 7.4.4.

Furthermore, the following **prerequisites** need to be satisfied:

(a) The new drill site shall be ready and complete in all respect before rig move is initiated.
(b) Mechanical and electrical crews are responsible for dismantling and fitting of pumps, engines, diesel tanks, electrical control room, and related work.
(c) Proper coordination among different crews/teams (e.g., drilling, mechanical, electrical, and mud) must be established. The crane crew is a critical resource during rig move operation; therefore, it is necessary that crane crews work in close coordination with drilling and other teams.
(d) Mud services group is responsible for the preparation of mud, and the time taken for preparation of spud mud varies from 6 to 18 h depending on the quality and type of mud used.

7.4.7 Model Execution

The critical path was computed and activity network was drawn using a computer. It was also computed manually that helps in understanding the process and its intricacies, and boosts one's confidence to solve such problem. The manual computation may be suitable for projects with small number of activities, but for big projects with large number of activities, manual computation and drawing of activity network are tedious work. The use of computer in such cases will be immensely beneficial.

The current critical path analysis model was also solved with the help of computer using one of the popular project management software packages. There are several software packages available in the market nowadays. One may choose a suitable one that would justify the need and cost and fit for the purpose. To learn more about network analysis, scheduling, PERT/CPM, and so on, the interested readers may refer to operations research and project management books listed under "References/Further Reading" at the end of this chapter.

7.5 RESULTS AND DISCUSSIONS

The results of the network model provide much information, which is discussed in detail in the following paragraphs:

(a) The critical path has been identified and shown in Appendix 7.C marked in light gray. These are indicated as follows:
   Rig-down: *Tasks #5-6-7-8-9-10-12-15-16-17-33*
   Rig-up: *Tasks #3-4-5-10-12-13-14-17-18-19-20-21-22-32*
   The day-wise/hour-wise schedule of activities is also depicted in Appendix 7.C.
(b) The **optimum rig move time is found as 5 days** (rig-down ~2 days, transportation ~½–1 day, and rig-up 2½ days) against the norm time of 10 days followed by the E&P company under study for such type and capacity of rig considering round-the-clock operation.

(c) Rig move operation in some organizations or countries is restricted to daytime and no work is permitted at night; thus, effective working hour per day is limited to 12 h. In such case, optimum rig move/mobilization time as determined earlier would be 10 days, instead of 5 days. There may be other variations in working hours during rig move followed by different companies or countries, such as round-the-clock operation during rig-down and rig-up, but transportation is restricted to daytime only. In such case, rig move time may increase by 1 or 2 days.

(d) Similar exercise may be conducted for other type and capacity of rigs (including high floor mast with substructure). Empirical estimation suggests that rig move norm time followed by the organization can be reduced to 4–10 days for different types of on-land rigs.

(e) Rig move time may be adversely affected, if the next drilling location is
  (i) at a faraway location from the current location (say, beyond 20 km) and/or
  (ii) the topography of the area possesses difficulties such as undulating terrain, hilly areas, forest, marshy land, and so on
  (iii) Rig move time may also be affected due to inclement weather such as freezing cold, poor visibility during winter, incessant rain during monsoon leading to slushy ground and waterlogging, and blistering heat in summer.
  (iv) It may also be affected in case there are narrow roads, weak bridges, high-tension power lines, and so on, falling in the route.
  (v) Apart from these, there may be day- or nighttime work restriction for certain activities (transportation, etc.).

Suitable *allowances may be considered for all such cases while fixing norm time.*

7.6 PROBABILITY OF COMPLETION OF PROJECT ON TIME

Managers and decision makers may be interested to know the probability of completion of project (i.e., rig move) within a specific date or time. They may also like to explore various options and examine the effect of varying project completion (rig move) time and related probability. There are two important aspects that are necessary conditions for determining the probability of completion of a project: First, the probability distribution function is required to be known or ascertained. Second, such computation is based on the following two assumptions:

(a) It is assumed that *activity times are independent random variables*. Therefore, the *project completion time that is summation of critical activity times* along the critical path *is also a random variable.*

(b) Project completion time (T_{cp}) that is a random variable tends to follow an approximated normal distribution. This assumption relies on central limit theorem of probability that specifies that the sum of n independent random variables tends to be normally distributed as n tends to infinity. Therefore, it may be assumed that if a critical path has many activities, then its completion time tends to follow approximated normal distribution.

Thus, as explained in (a), the length of the project or project completion time

$$T_{cp} = \sum T_{cp_i}$$

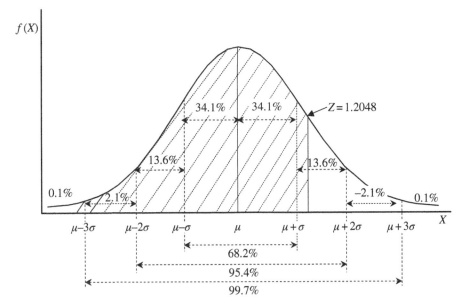

FIGURE 7.1 Normal distribution with mean (μ) and standard deviation (σ) depicting areas under each segment of the curve; shaded area left to the Z-value is the probability of occurrence.

Similarly, the project variance is the summation of individual independent activity times variance, that is, $\sigma = \Sigma \sigma_{cp_i}$, and standard deviation is the square root of variance, that is, $\sigma = (\sqrt{variance})$. With the help of Equations 7.1 and 7.2 (as explained in Section 7.4.3), project completion time (T_{cp}) and variance (σ^2) can be computed.

The higher the standard deviation, the greater the uncertainty that the project would be completed in time. It is quite likely that many of these critical activities may take longer than the expected time and there may be potential variability in the project completion time.

As explained in the assumption (b), the normal distribution is a bell-shaped curve that describes the statistical behavior of many real-world events. It is characterized by mean (μ) and standard deviation (σ) with widespread applications in real life. Figure 7.1 shows normal distribution curve that describes statistically significant conclusions, namely, a random variable drawn from a normal distribution has 68.26% chance of falling between $\mu \pm \sigma$, 95.44% chance of falling between $\mu \pm 2\sigma$, and 99.74% chance of falling between $\mu \pm 3\sigma$.

7.6.1 Computing Probability

Let T_{cp} be the project completion time following the critical path, which is essentially the mean project completion time (μ). Therefore, $T_{cp} = \mu$.

Let us find out the probability of completing the project within a specified time (say, T_{st}). For this, we need to convert T_{st} to a standard normal random variable Z as follows:

$$Z = \frac{T_{st} - T_{cp}}{\sigma} = \frac{T_{st} - \mu}{\sigma}; \rightarrow P\left(Z \le \frac{T_{st} - \mu}{\sigma}\right) \tag{7.3}$$

Referring to the standard normal distribution table, find out the value corresponding to $(T_{st} - \mu)/\sigma$, *which is the probability of completion of project within specific time* $P(Z \le T_{st})$.

For example, let the project (rig move) completion time be 5 days (as computed in Section 7.5b), that is, $T_{cp} = \mu = 5$ days.

Standard deviation for expected completion time $\sigma = 20$ h approximately 0.83 day (which is based on individual activity time estimate):

- What is the probability that rig move will be completed in 6 days?

$P(Z \le (6 - 5/0.83)) = P(Z \le 1.2048) = \textbf{89.4\%}$ (refer to Fig. 7.1: the area left to the z-value under the standard normal curve, that is, the area from the left-hand tail to specific point z, which is 1.2048 standard deviation above the mean). It signifies that the probability of completing rig move in 6 days is very high (89.4%).

- What is the probability that rig move will be completed in 4½ days?

$P(Z \le (4.5 - 5/0.83)) = P(Z \le -0.6024) = \textbf{26.63\%}$ (the area left to the z-value under the standard normal curve, that is, the area from the left-hand tail to specific point z, which is $(-)0.6024$ standard deviation below the mean). It signifies that the probability of completing rig move in 4½ days is not promising (26.63%).

7.7   CRASHING OF ACTIVITY AND PROJECT COMPLETION TIME

Critical path is the longest duration for completion of project using optimum (minimal) resources. Sometimes, it may be necessary to complete the project (i.e., rig move) earlier than the normal completion time due to exigency or other compelling reasons. *The process of reducing the normal activity or project completion time is called crashing.* This is usually done by deploying additional resources. For example, additional resources for crashing rig move schedule mean hiring of extra crew, payment of overtime, and hiring of cranes, trailers, and other equipment. But additional resources means higher cost; therefore, crashing of project leads to an increase in the activity/ project cost.

The project cost may be divided into three basic components, namely, the following:

(i) Direct cost of activities
(ii) Indirect cost that includes overhead cost, supervision, general administration, and related expenditure
(iii) Opportunity cost such as late penalty charges, late demurrages, loss of early completion incentives, and so on

Direct cost generally reduces with the increase in project completion time, but indirect and opportunity costs rise with the project length [6]. Figure 7.2 illustrates the time–cost relationship and characteristics of a project.

Optimal project completion time ensures minimum total project cost. Crashing cost generally increases with the shortening of project duration, as it moves away from the optimal project completion time. The project length may be shortened until optimal project completion time is reached.

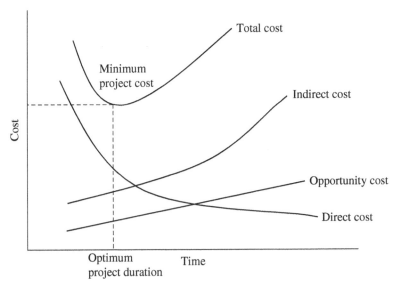

FIGURE 7.2 Project time–cost relationship.

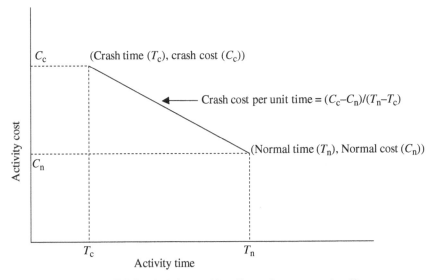

FIGURE 7.3 Activity crashing: linear time–cost trade-off.

Steps in Crashing

(a) Activities on the critical path are usually considered for crashing, as they influence project completion time. Generally, the least cost critical activity (LC_1) is preferred for crashing, followed by the next least cost critical activity (LC_2), and so on.

(b) Develop a time–cost table ranging from normal time to crash time and corresponding costs. Figure 7.3 depicts linear time–cost trade-off for crashing activities that are explained as follows:

Activity times are estimated based on normal working conditions. Normal time (T_n) is the maximum time required for completing an activity using normal or optimal resources. The cost required to complete an activity or project in normal time is called normal cost (C_n). Crash time (T_c) is the minimal possible time to complete an activity or project using maximum resources. The cost required to complete the project in crash time is called crash cost (C_c). The crashing of project leads to an increase in the project cost.

- For each activity, determine the cost of reduction (crashing) per unit time as follows:

- Crash cost per unit time $= \dfrac{\text{crash cost} - \text{normal cost}}{\text{normal time} - \text{crash time}} = \dfrac{C_c - C_n}{T_n - T_c}$.

(c) After every crashing, check the critical path, as it might change or there may be multiple critical paths. In such a situation, focus on common activities for further crashing.

(d) Repeat the process until minimum project duration is reached.

Crashing of activity or project may lower the quality of work/activity or may compromise with safety. Therefore, sufficient safeguards may be taken in this regard.

7.8 SUGGESTIONS

The recommendations of this study are summarized as follows:

(a) It is suggested to follow the critical path of rig move activity as determined and shown in Section 7.5a and in Appendix 7.C. *It is the optimum rig move schedule that would ensure minimum rig move time* for the type of rig under study.

(b) The sequence and duration of rig move activities may differ from one type of rig to another due to variation in design and capacity of rigs. Therefore, the schedule of rig move activities and rig move completion time of different types and capacity of rigs may be determined separately.

(c) Suitable allowances may be considered for standardization of rig move/mobilization time depending on the distance of rig relocation, difficult terrain and topography, road conditions, route and logistics issues, extreme weather conditions, day/night work restriction, and so on.

(d) In order to safeguard from the aforementioned adversities, adequate measures may be taken before the start of rig move operation that include **proper route planning and site reconnaissance**. Rig move team may develop detailed checklist including route survey, road obstacles and permits, coordination with local authorities, site conditions, and mobility constraints and ensure that these possess no hindrance [1].

(e) The company under study uses in-house resources for the entire rig move activity. Many E&P companies outsource rig move activities fully or partially depending on their capability and resource availability. **The decision whether rig move be done in-house or outsourced** *will depend on the availability of resources, capability of the company, and economic consideration. For the instant case, some activities* *like transportation and hiring of cranes may be outsourced for improved performance and faster rig mobilization.*

(f) **Safety aspects and safe operation** during rig move are of prime importance, as they involve dismantling, loading, transporting, unloading, and assembling of heavy equipment and machineries. It is necessary that E&P companies *develop and follow* **standard operating procedure (SOP)** *for rig move activity.* Rig move team may develop checklist to this effect and ensure that SOP is complied with and safety aspects are not compromised. *This would prevent damaging of heavy equipment and machineries, schedule delay, inefficiency, unsafe practices, and injuries.*

(g) The network model may be further refined by reviewing allocation of resources and duration of activities that are estimates only and examining further scope for reduction in rig mobilization time.

7.8.1 Benefits

The current rig move/mobilization norm time followed by the organization under study is 10 days for this type of rigs (1500/2000 HP desert electrical rig on wheels with fully disassembled units). If the new rig move schedule as determined by critical path analysis is followed, rig move time (as well as rig cycle time) would come down by 5 days. As explained in Section 7.5d, there is scope for reduction in norm time (4–10 days) for other types of rig.

Considering a modest reduction of 5 rig-days for each rig move activity and on an average that each rig completes 2 wells or 2 rig move activity in a year, there would be a savings of $2 \times 5 = 10$ rig-days per rig per year, which would be available for productive drilling operation. The company is having around 70 own on-land deep drilling rigs. It is estimated that out of these 70 rigs, such reduction in rig move time (5 rig-days) can be effected to at least 50 rigs, which are of similar type. Therefore, there is a scope for reduction in rig move (as well as rig cycle) time at 10 rig-days per rig per year $\times$ 50 rigs = 500 rig-days per year.

Considering average operating cost of on-land deep drilling rig at USD 60,000 per day, the annual savings would be approximately USD $60,000 \times 500$ rig-days = USD 30 million per year.

7.9 LIMITATIONS OF THE STUDY AND SCOPE FOR FURTHER WORK

PERT/CPM is a popular project management tool that is widely used across the industry. It is a relatively simple, easy-to-implement, and powerful technique, which has many advantages. Nonetheless, it has some limitations, which are mostly of pedagogical nature and interest. Some of these are summarized as follows:

(a) It is difficult to estimate crash cost and normal cost of each activity; one may also face difficulty in estimating normal time of different activities.

(b) The validity of assumptions that (i) activities are statistically independent, (ii) activity times follow β-distribution, and (iii) project completion time follows approximated normal distribution are sometimes questioned, even though they are found to be true in majority of the cases.

The network needs constant monitoring and update; in addition, special attention is required on critical path and critical activities, which are considered as extraneous work.

7.9.1 Resource Allocation under Constraints

The current network model is based on the assumptions that there is no constraint on resource availability and sufficient rig crew, cranes, and trailers are available for rig move activities. The validity of this assumption has been defended and explained in Section 7.4.4. But resources are limited and under varying circumstances, it may pose as constraints. In such a situation, it may be required to determine rig move time and activity schedule afresh.

The criteria for resource allocation under constraints are float, activity duration, and quantity of resource required. The procedures that are followed for allocation of scarce resources among competing activities at every halt are described as follows:

(i) Identify activities that are eligible for scheduling—that is, activities whose predecessors have been scheduled. It may be noted that an activity can start only after completion of all its predecessors and availability of required resources to perform the task.

(ii) Update ES and EF and determine float (EF − ES) of these activities that were not allocated at earlier halt time and also their succeeding activities.

Choose from these activities considering completion of immediate predecessors and resource availability.

(iii) Assign priorities based on float—the least float activity is given the highest priority followed by the next least float activity, and so on. Since the activities on the critical path are having zero float, critical activities are automatic choice for priority. In case of tie in floats, assign priority based on activity duration followed by quantum of resources required (activity that requires maximum resources).

(iv) Allocate resources one by one, sequentially as per the availability of resources.

(v) Plan for the next halt for resource allocation based on the availability of resources. Repeat the above steps until all activities are scheduled.

Nowadays, most of the network analysis problems are solved using computer, which is easy to handle and provides instant solution even for a large and multiple constraint problem. It is suggested that further work may be carried out by varying the resources (manpower) as provided in Appendix 7.A.

With the ready availability of powerful project management software packages in the market, such resource allocation and "what-if" tests can be done with relative ease. Different software packages have different features, customization, and processes for different functions of network model, but the commonality for resource allocation function includes the following:

• Specifying resource requirement for each task/activity (open resource sheet or create resource corresponding to each task/activity)

• Linking resources with the corresponding tasks/activities and following instructions for obtaining the results

7.9.2 Effect of Crashing

The effect of crashing on optimal rig move schedule may be studied by increasing/varying rig crew, cranes, and transport fleet and reducing completion time of critical activities. For this, crashing cost per unit time for each activity may be obtained or developed. This aspect could not be studied in this exercise for lack of related cost data/information. With the availability of such data, this aspect can be studied following the steps as explained in Section 7.7.

7.10 CONCLUSIONS

Rig rental and operating cost of deep drilling rig is very high and cost of drilling is continuously rising. Therefore, E&P companies and drilling operators are looking for all possible measures and avenues to minimize unproductive rig time and control soaring drilling cost. Network analysis using PERT/CPM technique has been used in this study to optimize rig move/mobilization time and develop optimal schedule of rig move activity.

The critical path of rig move/mobilization was determined and critical activities were identified by computing earliest start ES, EF, LS, LF, and float of each activity. The computation was done by making forward passes as well as backward passes through the network. PERT/CPM technique also helps in decision making and "what-if" test, finding alternative schedule, determining the effect of varying resources, and crashing activities on project completion time.

Finally, a set of **suggestions** have been made for improving the efficiency of rig mobilization, which are summarized as follows:

(a) It is suggested to *follow the critical path* of rig move activity as determined and shown in Appendix 7.C and in Section 7.5a. *The critical path ensures optimal scheduling of activities and rig move time, which is found as 5 days* (rig-down ~2 days, transportation ~½–1 day, rig-up ~2½ days) against the norm time of 10 days followed by the E&P company under study. There is considerable scope for reduction in rig move norm time for other types of rigs.

(b) The sequence and duration of rig move activity depend on the type, capacity, and design of rigs. Therefore, rig move time and critical path for different types of rigs may be determined separately.

(c) Rig mobilization time is also affected by the distance of relocation, difficult terrain and topography, road conditions, route (weak bridges and high-tension power lines), extreme climate, day- or nighttime restrictions for certain activities, and so on, and these factors may be taken into account while standardizing the rig move/ mobilization time.

(d) Proper route planning and site reconnaissance activities may be undertaken before the rig move operation starts. Rig move team must check and ensure that road obstacles and permits, site conditions and mobility constraints, coordination with various agencies and authorities, and so on, possess no hindrance.

(e) E&P companies across the world use both in-house resources and outsourcing of rig move activity depending on the capability of the company, availability of resources, and economic considerations. For the present case, the use of both in-house resources and partial outsourcing of rig move activities such as transportation, hiring of cranes, and so on, will ensure improved performance and faster rig mobilization.

(f) **SOP** for rig move activities may be developed and followed that would ensure safe operations. It would prevent damaging of heavy machineries and critical equipment, schedule delay, inefficiency, and unsafe practices.

(g) Further scope for reduction of rig move time may be explored by reviewing duration of critical tasks, allocation of resources, effect of crashing, and so on.

7.10.1 Benefits

The existing rig move norm time for 1500/2000 HP desert electrical rig on wheels with fully disassembled units can be reduced to 5 days by following rig move schedule as computed through critical path analysis. Considering the number of rig move activities that actually takes place in onshore Assets/Basins of the company, there is a scope of saving of 500 rig-days amounting to USD 30 million per annum.

In Chapter 8, we would focus on some safety aspects, which are often neglected at offshore installations. It deals with developing uniform standard for emergency alarm system and code of signals for offshore installations, which are found to vary widely creating confusion among offshore going personnel.

REVIEW EXERCISES

7.1 What are the main components of rig move/mobilization activity? Why it is important for E&P companies to optimize the rig move time?

7.2 What are the factors that generally affect rig move time? What are the ways to control or mitigate these?

7.3 Distinguish the difference between PERT and CPM. Give some examples of their applicability. Where PERT is more suitable than CPM and vice versa?

7.4 How would you develop a network analysis model? Explain the steps involved.

7.5 Describe the procedures for determining critical path, slack, or float?

7.6 Define the following terms:
   (a) Critical path
   (b) Critical activity
   (c) Three time estimates
   (d) Earliest start
   (e) Latest start
   (f) Earliest finish
   (g) Latest finish
   (h) Float

7.7 What is dummy activity and what is its significance? Why it is used in a network?

7.8 How would you compute the probability of completing a project within a specified time (T_{st}) if mean project completion time (T_{cp}), standard deviation (σ) of project completion time, and standard normal distribution curve are available?

7.9 As a network specialist, what are the steps you would suggest for resource allocation under constraint?

7.10 What is "crashing" in network technique? As a project manager, you have been asked to crash a project—describe the procedure you would follow.

7.11 What are the different types of cost associated with network technique that you would focus on as a project manager?

APPENDIX 7.A   LIST OF TASKS/ACTIVITIES

Code	Task/Activity	Duration (h)	Predecessor	Manpower Required
Rig-Down				
	Rigging down	σ	σ	σ
	Substructure	σ	σ	σ
3	Dismantle H-manifold/S-bends; hook up bull line to draw-work power line pulley	6		6
4	Disengaging draw-work chain from rotary table	2	3,16	6
5	Check jacking and check water fill in the bottom box; disengage draw-work skid from rotary table skid; lower draw works	5	15,4	6
6	Removal of foundation bolts and R/Dn draw works	4	17	6
7	Dismantle setback and front spreader, Z-frame, and draw-work skid and spreader	6	6	6
8	Lowering of A-frames	2	7	6
9	Dismantle "A"-frame	4	8	
10	Emptying water from the bottom box, followed by the removal of bottom box extensions and jacking down and putting substructure on wheels	4	9	4
	Mast			
12	Rig-down tongs; remove all handling tools and sensors and lines	4		4
13	Dismantle topman emergency escape device; folding of stabbing board and fingers on monkey board	2		2
14	Winding up of winch lines and removal of winch and securing all other hanging lines	4	15,16	2
15	Removal of railings, platforms, and ladders; Rig-down doghouse and base; removal of catwalk	4	16	6
16	Remove TDS and swivel	8	12,13	4
17	Shift the position of bull line; lower the mast and place it on big horse	4	5,14,32	4
18	Dismantle monkey board and belly board, followed by resting of mast on small horse safely	2	17	6

(Continued)

Code	Task/Activity	Duration (h)	Predecessor	Manpower Required
19	Removal of ladders and topman's rest platform and winding of all hanging lines	2	20	6
20	Disengage bull line from yoke	2	18	6
21	De-reeving of drilling line and winding up in spool	3	18	6
22	Dismantle crown block and mast	6	21	6
23	Removal of all power lines	4	19	8
	Other accessories			
25	Disengage BOP control unit	2		2
26	Dismantle all cabins	4		4
27	Dismantle mud tanks, desander, desilter, and shale shaker	6	23	4
28	Dismantle water tanks and diesel tanks	4	27	4
29	Disengage all HP lines	4	3	4
30	Dismantle all engines and control room	6	17	
31	Dismantle and remove mud pumps and accessories	6	27	
32	Removal of tubulars, pipe rack, cable trays, and so on	4		
33	Putting on wheels for rig move	5	25,26,27, 28,30, 31,10	4

Rig-Up

Rigging Up

Substructure

Code	Task/Activity	Duration (h)	Predecessor	Manpower Required
3	Placement and alignment of substructure and erection of strong back on base; fill water in bottom box; removing of wheels and jacking	6		4
4	Connection of rear and front floor to the base	6	3	4
5	Connection of mast to the front floor	2	4,16	4
6	Fixing of draw works, rotary drive, and motor with rear floor; connect spreader beams and control panel and its connnections	6	4	4
7	Assembling of "A"-frame	8	4	4
8	Swiving of "A"-frame	3	7	4
9	Install doghouse, derrick platform, handrails, and so on	4	21	4
10	Fix air winch, rotary table, tongs, and tensors	4	21	4
11	Fitting of "S"-bends, rotary chain, drillometer, and conductor with shale shaker	6	10	4
12	Fitting of swivel, kelly bush, rotary hose with swivel, power tong, and so on	5	10	6
13	Install TDS	10	12	4
14	Function test TDS	1	13,22	2
	Mast			
16	Assemble mast pieces	4		6
17	Fix crown block, ladder, and tong counterweight lines	4	5	6

Code	Task/Activity	Duration (h)	Predecessor	Manpower Required
18	Fix belly board and monkey board, crown safety platform with handrails, boom for cathead line, standpipe, rotary hose, electric fittings, and other accessories	6	17	6
19	Reeving of casing line between crown block and traveling block and putting traveling block at the center of the mast; put bull line in mast lifting pulley	5	18	6
20	Raising of mast	2	8,19,28,6	4
21	Put bull line in draw-work lifting pulley; raise draw work	4	20	
22	Mast alignment and centering	2	21,13	
	Other accessories			
24	Placing of mud pumps and supercharger and their alignments	6	25	4
25	Placing of mud tanks and their connections	6		4
26	Placing of hopper, water tanks, and connections	6	24,25	2
27	Fixing of other pumps	6	24,25	4
28	Alignment of engines, diesel tanks, and elect. control room	12		2
29	Fix air lines, cable trays, adjustable flight stairways, catwalk and slope, pipe racks, tubulars, and so on	6	9,11,20	4
30	Preparation of mud	18	26,27	2
31	Install and function test of BOP unit	1	29	2
32	Testing of circulation system and spud of well	1	14,22,30,31	2

APPENDIX 7.B SLACK

Task No.	Task Name	Start	Finish	Late Start	Late Finish	Free Slack	Total Slack
	Rig-Down						
1	Rigging down	Tue 11/5/13	Thu 11/5/13	Tue 11/5/13	Thu 11/5/13	0 days	0 days
2	Substructure	Tue 11/5/13	Thu 11/5/13	Tue 11/5/13	Thu 11/5/13	0 days	0 days
3	Dismantle H-manifold/S-bends; hook up bull line to draw-work power line pulley	Tue 11/5/13	Tue 11/5/13	Tue 11/5/13	Tue 11/5/13	0h	8h
4	Disengaging draw-work chain from rotary table	Tue 11/5/13	Tue 11/5/13	Thu 11/5/13	Thu 11/5/13	32h	32h
5	Check jacking and check water fill in bottom box; disengage draw-work skid from rotary table skid; lower draw works	Tue 11/5/13	Wed 11/5/13	Tue 11/5/13	Wed 11/5/13	0h	0h
6	Removal of foundation bolts and R/Dn draw works	Wed 11/6/13	Wed 11/6/13	Wed 11/6/13	Wed 11/6/13	0h	0h
7	Dismantle Z-frame and draw-work skid and spreader	Wed 11/6/13	Wed 11/6/13	Wed 11/6/13	Wed 11/6/13	0h	0h
8	Lowering of A-frames	Wed 11/6/13	Wed 11/6/13	Wed 11/6/13	Wed 11/6/13	0h	0h
9	Dismantle "A"-frame	Wed 11/6/13	Thu 11/7/13	Wed 11/6/13	Thu 11/7/13	0h	0h
10	Emptying water from bottom box, followed by removal of bottom box extensions and jacking down and putting substructure on wheels	Thu 11/7/13	Thu 11/7/13	Thu 11/7/13	Thu 11/7/13	0h	0h
11	Mast	Tue 11/5/13	Wed 11/6/13	Tue 11/5/13	Thu 11/7/13	0 days	0 days
12	Rig-down tongs; remove all handling tools and sensors and lines	Tue 11/5/13	Tue 11/5/13	Tue 11/5/13	Tue 11/5/13	0h	0h
13	Dismantle topman emergency escape device; folding of stabbing board and fingers on monkey board	Tue 11/5/13	Tue 11/5/13	Tue 11/5/13	Tue 11/5/13	2h	2h
14	Winding up of winch lines and removal of winch and securing all other hanging lines	Tue 11/5/13	Wed 11/6/13	Thu 11/7/13	Thu 11/7/13	26h	26h
15	Removal of railings, platforms, and ladders; rig-down doghouse and base; removal of catwalk	Tue 11/5/13	Tue 11/5/13	Tue 11/5/13	Tue 11/5/13	0h	0h
16	Remove TDS and swivel	Tue 11/5/13	Tue 11/5/13	Tue 11/5/13	Tue 11/5/13	0h	0h
17	Shift the position of bull line; lower the mast and place it on big horse	Wed 11/6/13	Wed 11/6/13	Wed 11/6/13	Wed 11/6/13	0h	0h
18	Dismantle monkey board and belly board, followed by resting of mast on small horse safely	Wed 11/6/13	Wed 11/6/13	Wed 11/6/13	Wed 11/6/13	0h	6h
19	Removal of ladders and topman's rest platform and winding of all hanging lines	Wed 11/6/13	Wed 11/6/13	Wed 11/6/13	Wed 11/6/13	0h	6h
20	Disengage bull line from yoke	Wed 11/6/13	Wed 11/6/13	Wed 11/6/13	Wed 11/6/13	0h	6h

21	De-reeving of drilling line and winding up in spool	Wed 11/6/13	Wed 11/6/13	Thu 11/7/13	Thu 11/7/13	0h	14h
22	Dismantle crown block and mast	Wed 11/6/13	Wed 11/6/13	Thu 11/7/13	Thu 11/7/13	14h	14h
23	Removal of all power lines	Wed 11/6/13	Wed 11/6/13	Wed 11/6/13	Thu 11/7/13	0h	6h
24	Other accessories	Tue 11/5/13	Tue 11/5/13	Wed 11/6/13	Thu 11/7/13	0 days	0 days
25	Disengage BOP control unit	Tue 11/5/13	Tue 11/5/13	Tue 11/5/13	Thu 11/7/13	44h	44h
26	Dismantle all cabins	Wed 11/6/13	Wed 11/6/13	Wed 11/6/13	Thu 11/7/13	6h	6h
27	Dismantle mud tanks, desander, desilter, and shale shaker	Tue 11/5/13	Tue 11/5/13	Wed 11/6/13	Wed 11/6/13	0h	34h
28	Dismantle water tanks and diesel tanks	Tue 11/5/13	Tue 11/5/13	Thu 11/7/13	Thu 11/7/13	36h	36h
29	Disengage all HP lines	Tue 11/5/13	Tue 11/5/13	Thu 11/7/13	Thu 11/7/13	36h	36h
30	Dismantle all engines and control room	Wed 11/6/13	Wed 11/6/13	Wed 11/6/13	Thu 11/7/13	14h	14h
31	Dismantle and remove mud pumps and accessories	Tue 11/5/13	Tue 11/5/13	Thu 11/7/13	Thu 11/7/13	34h	34h
32	Removal of tubulars, pipe rack, cable trays, and so on	Tue 11/5/13	Tue 11/5/13	Thu 11/7/13	Thu 11/7/13	42h	42h
33	Putting on wheels for rig move	Thu 11/7/13	Thu 11/7/13	Thu 11/7/13	Thu 11/7/13	0h	0h

Rig-Up

1	Rigging up	Fri 11/8/13	Sun 11/10/13	Fri 11/8/13	Sun 11/10/13	0 days	0 days
2	Substructure	Fri 11/8/13	Sun 11/10/13	Fri 11/8/13	Sun 11/10/13	0 days	0 days
3	Placement and alignment of substructure and erection of strong back on base; fill water in bottom box; removing wheels and jacking	Fri 11/8/13	Fri 11/8/13	Fri 11/8/13	Fri 11/8/13	0h	0h
4	Connection of rear and front floor to the base	Fri 11/8/13	Fri 11/8/13	Fri 11/8/13	Fri 11/8/13	0h	0h
5	Connection of mast to the front floor	Fri 11/8/13	Fri 11/8/13	Fri 11/8/13	Fri 11/8/13	0h	0h
6	Fixing of draw works, rotary drive, and motor with rear floor; connect spreader beams and control panel and its connections	Fri 11/8/13	Fri 11/8/13	Sun 11/10/13	Sun 11/10/13	39h	39h
7	Assembling of "A"-frame	Fri 11/8/13	Fri 11/8/13	Fri 11/8/13	Sat 11/8/13	0h	6h
8	Swiving of "A"-frame	Fri 11/8/13	Fri 11/8/13	Sat 11/9/13	Sat 11/9/13	6h	6h
9	Install doghouse, derrick platform, handrails, and so on	Sat 11/9/13	Sat 11/9/13	Sun 11/10/13	Sun 11/10/13	18h	18h
10	Fix air winch, rotary table, tongs, and tensors	Sat 11/9/13	Sat 11/9/13	Sat 11/9/13	Sat 11/9/13	0h	0h
11	Fitting of "S"-bends, rotary chain, drillometer, and conductor with shale shaker	Sat 11/9/13	Sat 11/9/13	Sun 11/10/13	Sun 11/10/13	12h	12h
12	Fitting of swivel, kelly bush, rotary hose with swivel, power tong, and so on	Sat 11/9/13	Sat 11/9/13	Sat 11/9/13	Sat 11/9/13	0h	0h
13	Install TDS	Sat 11/9/13	Sat 11/9/13	Sat 11/9/13	Sun 11/10/13	0h	0h

(Continued)

Task No.	Task Name	Start	Finish	Late Start	Late Finish	Free Slack	Total Slack
14	Function test TDS	Sun 11/10/13	Sun 11/10/13	Sun 11/10/13	Sun 11/10/13	0h	0h
15	Mast	Fri 11/8/13	Sun 11/10/13	Fri 11/8/13	Sun 11/10/13	0 days	0 days
16	Assemble mast pieces	Fri 11/8/13	Fri 11/8/13	Fri 11/8/13	Fri 11/8/13	8h	8h
17	Fix crown block, ladder, and tong counterweight lines	Fri 11/8/13	Fri 11/8/13	Fri 11/8/13	Fri 11/8/13	0h	0h
18	Fix belly board and monkey board, crown safety platform with handrails, boom for cathead line, standpipe, rotary hose, electric fittings, and other accessories	Fri 11/8/13	Sat 11/9/13	Fri 11/8/13	Sat 11/9/13	0h	0h
19	Reeving of casting line between crown block and traveling block and put traveling block at the center of the mast; put bull line in mast lifting pulley	Sat 11/9/13	Sat 11/9/13	Sat 11/9/13	Sat 11/9/13	0h	0h
20	Raising of mast	Sat 11/9/13	Sat 11/9/13	Sat 11/9/13	Sat 11/9/13	0h	0h
21	Put bull line in draw-work lifting pulley; raise draw work	Sat 11/9/13	Sat 11/9/13	Sat 11/9/13	Sat 11/9/13	0h	0h
22	Mast alignment and centering	Sun 11/10/13	Sun 11/10/13	Sun 11/10/13	Sun 11/10/13	0h	0h
23	Other accessories	Fri 11/8/13	Sun 11/10/13	Fri 11/8/13	Sun 11/10/13	0 days	0 days
24	Placing of mud pumps and supercharger and their alignments	Fri 11/8/13	Fri 11/8/13	Sat 11/9/13	Sat 11/9/13	0h	21h
25	Placing of mud tanks and their connection	Fri 11/8/13	Fri 11/8/13	Sat 11/9/13	Sat 11/9/13	0h	21h
26	Placing of hopper, water tanks, and connections	Fri 11/8/13	Fri 11/8/13	Sat 11/9/13	Sat 11/9/13	0h	21h
27	Fixing of other pumps	Fri 11/8/13	Fri 11/8/13	Sat 11/9/13	Sat 11/9/13	0h	21h
28	Alignment of engines, diesel, tanks, and elect. control room	Fri 11/8/13	Fri 11/8/13	Sat 11/9/13	Sun 11/10/13	45h	45h
29	Fix air lines, cable trays, adjustable flight stairways, catwalk and slope, pipe racks, tubulars, and so on	Sat 11/9/13	Sun 11/10/13	Sun 11/10/13	Sun 11/10/13	0h	5h
30	Preparation of mud	Fri 11/8/13	Sat 11/9/13	Sat 11/9/13	Sun 11/10/13	21h	21h
31	Install and function test of BOP unit	Sun 11/10/13	Sun 11/10/13	Sun 11/10/13	Sun 11/10/13	5h	5h
32	Testing of circulation system and spud of well	Sun 11/10/13	Sun 11/10/13	Sun 11/10/13	Sun 11/10/13	0h	0h

APPENDIX 7.C CRITICAL PATH

Ring-Down

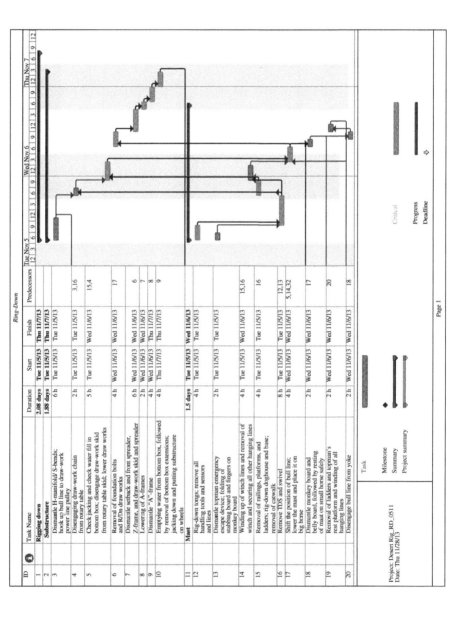

ID	❶	Task Name	Duration	Start	Finish	Predecessors
1		**Rigging down**	**2.08 days**	**Tue 11/5/13**	**Thu 11/7/13**	
2		**Substructure**	**1.88 days**	**Tue 11/5/13**	**Thu 11/7/13**	
3		Dismantle H-manifold/ S-bends; hook up bull line to draw-work power line pulley	6 h	Tue 11/5/13	Tue 11/5/13	
4		Disengaging draw-work chain from rotary table	2 h	Tue 11/5/13	Tue 11/5/13	3,16
5		Check jacking and check water fill in bottom box; disengage draw-work skid from rotary table skid; lower draw works	5 h	Tue 11/5/13	Wed 11/6/13	15,4
6		Removal of foundation bolts and R/Dn draw works	4 h	Wed 11/6/13	Wed 11/6/13	17
7		Dismantle setback and front spreader, Z-frame, and draw-work skid and spreader	6 h	Wed 11/6/13	Wed 11/6/13	6
8		Lowering of A-frames	2 h	Wed 11/6/13	Wed 11/6/13	7
9		Dismantle "A"-frame	4 h	Wed 11/6/13	Thu 11/7/13	8
10		Emptying water from bottom box, followed by removal of bottom box extensions; jacking down and putting substructure on wheels	4 h	Thu 11/7/13	Thu 11/7/13	9
11		**Mast**	**1.5 days**	**Tue 11/5/13**	**Wed 11/6/13**	
12		Rig-down tongs; remove all handling tools and sensors and lines	4 h	Tue 11/5/13	Tue 11/5/13	
13		Dismantle topman emergency escape device; folding of stabbing board and fingers on monkey board	2 h	Tue 11/5/13	Tue 11/5/13	
14		Winding up of winch lines and removal of winch and securing all other hanging lines	4 h	Tue 11/5/13	Wed 11/6/13	15,16
15		Removal of railings, platforms, and ladders; rig-down doghouse and base; removal of catwalk	4 h	Tue 11/5/13	Tue 11/5/13	16
16		Remove TDS and swivel	8 h	Tue 11/5/13	Tue 11/5/13	12,13
17		Shift the position of bull line; lower the mast and place it on big horse	4 h	Wed 11/6/13	Wed 11/6/13	5,14,32
18		Dismantle monkey board and belly board, followed by resting of mast on small horse safely	2 h	Wed 11/6/13	Wed 11/6/13	17
19		Removal of ladders and topman's rest platform and winding of all hanging lines	2 h	Wed 11/6/13	Wed 11/6/13	20
20		Disengage bull line from yoke	2 h	Wed 11/6/13	Wed 11/6/13	18

Project: Desert Rig_RD_0511
Date: Thu 11/28/13

| Task | ▬▬▬ | Milestone | ◆ |
| Summary | | Project summary | |

Critical ▭▭▭

Progress ▬▬▬
Deadline ⇩

Page 1

ID	ⓘ	Task Name	Duration	Start	Finish	Predecessors
21		De-reeving of drilling line and winding up in spool	3 h	Wed 11/6/13	Wed 11/6/13	18
22		Dismantle crown block and mast	6 h	Wed 11/6/13	Wed 11/6/13	21
23		Removal of all power lines	4 h	Wed 11/6/13	Wed 11/6/13	19
24		**Other accessories**	**2.08 days**	**Tue 11/5/13**	**Thu 11/7/13**	
25		Disengage BOP control unit	2 h	Tue 11/5/13	Tue 11/5/13	
26		Dismantle all cabins	4 h	Wed 11/6/13	Wed 11/6/13	23
27		Dismantle mud tanks, desander, desilter, and shale shaker	6 h	Tue 11/5/13	Tue 11/5/13	
28		Dismantle water tanks and diesel tanks	4 h	Tue 11/5/13	Tue 11/5/13	27
29		Disengage all HP lines	4 h	Tue 11/5/13	Tue 11/5/13	3
30		Dismantle all engines, control room	6 h	Wed 11/6/13	Wed 11/6/13	17
31		Dismantle and remove mud pumps and accessories	6 h	Tue 11/5/13	Tue 11/5/13	27
32		Removal of tubulars, pipe rack cable trays, and so on	4 h	Tue 11/5/13	Tue 11/5/13	
33		Putting on wheels for rig move	5 h	Thu 11/7/13	Thu 11/7/13	25,26,27,28,30,31,10

Task

Milestone

Summary

Project summary

Critical

Progress

Deadline

Ring-Up

ID	ⓘ	Task Name	Duration	Start	Finish	Predecessors	Fri Nov 8	Sat Nov 9	Sun Nov 10
							6 9 12 3 6 9 12 3 6 9	12 3 6 9 12 3 6 9 12 3 6 9	12 3 6 9 12
1		**Rigging up**	**2.42 days**	**Fri 11/8/13**	**Sun 11/10/13**				
2		**Substructure**	**2.38 days**	**Fri 11/8/13**	**Sun 11/10/13**				
3		Placement and alignment of substructure and erection of strong back on base; fill water in bottom box; removing wheels and jacking	6 h	Fri 11/8/13	Fri 11/8/13				
4		Connection of rear and front floor to the base	6 h	Fri 11/8/13	Fri 11/8/13	3			
5		Connection of mast to the front floor	2 h	Fri 11/8/13	Fri 11/8/13	4,16			
6		Fixing of draw works, rotary drive, and motor with rear floor; connect spreader beams and control panel and its connnections	6 h	Fri 11/8/13	Fri 11/8/13	4			
7		Assembling of "A"-frame	8 h	Fri 11/8/13	Fri 11/8/13	4			
8		Swiving of "A"-frame	3 h	Fri 11/8/13	Fri 11/8/13	7			
9		Install doghouse, derrick platform, handrails, and so on	4 h	Sat 11/9/13	Sat 11/9/13	21			
10		Fix air winch, rotary table, tongs, and tensors	4 h	Sat 11/9/13	Sat 11/9/13	21			
11		Fitting of "S"-bends, rotary chain, drillometer, and conductor with shale shaker	6 h	Sat 11/9/13	Sat 11/9/13	10			
12		Fitting of swivel, kelly bush, rotary hose with swivel, power tong, and so on	5 h	Sat 11/9/13	Sat 11/9/13	10			
13		Install TDS	10 h	Sat 11/9/13	Sun 11/10/13	12			
14		Function test TDS	1 h	Sun 11/10/13	Sun 11/10/13	22,13			
15		**Mast**	**2.33 days**	**Fri 11/8/13**	**Sun 11/10/13**				
16		Assemble mast pieces	4 h	Fri 11/8/13	Fri 11/8/13				
17		Fix crown block, ladder, and tong counter weight lines	4 h	Fri 11/8/13	Fri 11/8/13	5			
18		Fix belly board and monkey board, crown safety platform with hand rails, boom for cathead line, standpipe, rotary hose, electric fittings, and other accessories	6 h	Fri 11/8/13	Sat 11/9/13	17			

Project: Rig up_DR_0511
Date: Sun 12/1/13

Task	
Milestone	◆
Summary	
Project summary	

Critical	
Progress	
Deadline	⇩

Page 1

ID	O	Task Name	Duration	Start	Finish	Predecessors
19		Reeving of casing line between crown block and traveling block and put traveling block at the center of the mast; put bull line in mast lifting pulley	5 h	Sat 11/9/13	Sat 11/9/13	18
20		Raising of mast	2 h	Sat 11/9/13	Sat 11/9/13	8,19,28,6
21		Put bull line in draw-work lifting pulley; raise draw work	4 h	Sat 11/9/13	Sat 11/9/13	20
22		Mast alignment and centering	2 h	Sun 11/10/13	Sun 11/10/13	21,13
23		**Other Accessories**	**2.42 days**	**Fri 11/8/13**	**Sun 11/10/13**	
24		Placing of mud pumps and supercharger and their alignments	6 h	Fri 11/8/13	Fri 11/8/13	25
25		Placing of mud tanks and their connections	6 h	Fri 11/8/13	Fri 11/8/13	
26		Placing of hopper, water tanks, and connections	6 h	Fri 11/8/13	Fri 11/8/13	24,25
27		Fixing of other pumps	6 h	Fri 11/8/13	Fri 11/8/13	24,25
28		Alignment of engines, diesel tanks, and elect. control room	12 h	Fri 11/8/13	Fri 11/8/13	
29		Fix air lines, cable trays, adjustable flight stair ways, catwalk and slope, pipe racks, tubulars, and so on	6 h	Sat 11/9/13	Sun 11/10/13	20,9,11
30		Preparation of mud	18 h	Fri 11/8/13	Sat 11/9/13	26,27
31		Install and function test BOP unit	1 h	Sun 11/10/13	Sun 11/10/13	29
32		Testing of circulation system and spud of well	1 h	Sun 11/10/13	Sun 11/10/1	14,22,31,30

Task		Critical	
Milestone		Progress	
Summary		Deadline	
Project summary			

REFERENCES AND USEFUL LINKS

[1] Buehler, A. and Charwinsky, J., Who Moved My Rig? Oil and Gas Financial Journal, December 2011, available at: http://www.ogfj.com/articles/print/volume-8/issue-12/features/who-moved-my-rig.html (accessed on March 9, 2016).

[2] Burk, V. and Fink, M., Utilizing Lean Concepts in E&P: The Time for Lean Is Now, A Study Conducted by Alvarez & Marsal (A&M) Services, available at: http://www.alvarezandmarsal.com/sites/default/files/sidebar-callouts/bc_-_utilizing_lean_concepts_w-p_1.pdf (accessed on March 9, 2016).

[3] Wagner, H.M., Principles of Operations Research, Prentice-Hall of India, New Delhi, 1980.

[4] Weist, J.D. and Levy, F.K., A Management Guide to PERT/CPM, Prentice-Hall of India Private Ltd., New Delhi, 1982.

[5] Project Management: PERT and CPM (Chapter 14), Business Modeling and Simulation course, Department of Operations Management and Information Systems, KIMEP University, Pr. Abaya, Kazakhstan, available at: http://www2.kimep.kz/bcb/omis/our_courses/is4201/Chap14.pdf (accessed on March 9, 2016).

[6] McKenna, C.K., Quantitative Methods for Public Decision Making, McGraw-Hill, New York, 1980.

FURTHER READING

Anderson, D.R., Sweeney, D.J., and Williams, T.A., An Introduction to Management Science: Quantitative Approaches to Decision Making, 7th edition, West Publishing Company, Minneapolis, MN, 1994.

Archibald, R.D. and Villoria, R.L., Network-Based Management Systems (PERT/CPM), John Wiley & Sons, Inc., New York, 1967.

Clayton, R., White, M.J., and Myrtle, R., Managing Public Systems: Analytic Techniques for Public Administration, revised edition, University Press of America, Lanham, MD, 1985.

Cleland, D.I. and King, W., Systems Analysis and Project Management, 2nd edition, McGraw-Hill, New York, 1975.

Evarts, H.G., Introduction to PERT, Allyn and Bacon, Boston, MA, 1964.

Kerzner, H., Project Management: A Systems Approach to Planning, Scheduling, and Controlling, 10th edition, John Wiley & Sons, Inc., Hoboken, NJ, 2009.

Levin, R.I. and Kirkpatrick, C.A., Planning and Control with PERT/CPM, McGraw-Hill, New York, 1966.

Meredith, J.R. and Mantel, S.J., Jr., Project Management: A Managerial Approach, 7th edition, John Wiley & Sons, Inc., Hoboken, NJ, 2008.

Moder, J.J. and Phillips, C.R., Project Management with CPM and PERT, Van Nostrand Reinhold Company, New York, 1970.

Morris, L.N., Critical Path: Construction and Analysis, Pergamon Press, New York, 1967.

Weiss, J. and Wysocki, R., Five-phase Project Management: A Practical Planning and Implementation Guide, Basic Books, New York, 1992.

Whitehouse, G.E., Systems Analysis and Designs using Network Techniques, Prentice-Hall Inc., Englewood Cliffs, NJ, 1970.

Wisniewski, M. and Klein, J.H., Critical Path Analysis and Linear Programming, Palgrave Macmillan, Basingstoke, 2001.

Useful Links

Hiemstra, R., Critical Path Analysis: A Planning/Time Management Tool for Managing Research, 2000, available at: http://roghiemstra.com/cpa.html (accessed on March 9, 2016).

PERT/CPM for Development Project Scheduling & Management, Intervention (India) Pvt. Ltd., Bangalore, available at: http://www.interventions.org/pertcpm.html (accessed on March 9, 2016).

Project Scheduling Models (Chapter 5), Training activities G Staff input on prototype models, National Taipei University, Taiwan, available at: web.ntpu.edu.tw/~juang/ms/CH05.ppt (accessed on March 9, 2016).

Clayton, R., School of Public Administration, University of Southern California, Los Angeles, Techniques of Network Analysis for Managers (Chapter 3), available at: web.pdx.edu/~stipakb/download/PA557/PERT-Overview2.pdf (accessed on March 9, 2016).

8

DEVELOPING UNIFORM STANDARDS FOR EMERGENCY ALARM SYSTEM AND INDICATORS FOR OFFSHORE OIL AND GAS INSTALLATIONS

8.1 INTRODUCTION

We have dealt with issues related to core business activities and critical functional areas in the previous chapters. In this chapter, we would deal with an important functional area, namely, health, safety, and environment (HSE), which is often overlooked or neglected for short-term gain. HSE has gained importance in today's business world in general and oil and gas industry in particular. Gone are those days when safety department (HSE department was nonexistent at that time) in an organization was perceived as less important, and posting there was considered not so glorious! With the growing concern for health and safety of employees, rising public awareness, and enforcement of stringent rules for environment, HSE department has gained unprecedented importance these days all over the world.

Nowadays, HSE is as important a function as production and operations. It is no longer a social obligation but an imperative for business continuity, especially in oil and gas industry. It is a well-known fact that the lack of safety culture, negligence of health, and safety of employees in an enterprise often leads to accident, loss of life, and property. These, in turn, adversely affect employee morale, productivity, and reputation of the company. Therefore, it is necessary to take appropriate safeguards and measures for the betterment of health and safety in an organization. Similarly, environmental degradation caused by industrial units (oil and gas installations) may prove dear in the wake of rising public awareness and stringent environmental laws. Therefore, it is important to follow the code of safe practices and imbibe the safety culture in the organization, especially in oil and gas industry. It assumes greater importance for offshore E&P operations, as

Optimization and Business Improvement Studies in Upstream Oil and Gas Industry, First Edition.
Sanjib Chowdhury.
© 2016 John Wiley & Sons, Inc. Published 2016 by John Wiley & Sons, Inc.

it is associated with high degree of risks and requires extreme caution at every step. *Despite such vulnerable and high-stake operations, many seemingly innocuous issues escape attention. The current study deals with one such issue, which apparently looks inconsequential but may prove savior of life and property during emergency.*

8.1.1   Background of the Problem

This study was carried out for offshore Assets of an E&P company, which are spread over a vast area. These offshore Assets have numerous rigs, platforms, and installations, which are manned round the clock. Offshore operations are strenuous, stressful, and risky; therefore, it is important that crew members follow the safe code of practices at every step and are alert all the time.

It was observed in this study that emergency alarm systems and indicators used at various offshore rigs, platforms, and installations are not the same. They differ from one installation to another even in the same Asset, not to speak of across Assets. This creates confusion among crew members, contractor's personnel, and others who frequent different rigs and installations for special tasks and jobs. This issue, though looks innocuous, has damaging consequences if not redressed.

It is of paramount importance to bring uniformity in code of signals and standards for emergency alarm and indicators for offshore installations, so that there is no room for confusion. It would be advantageous if uniform code of signals is followed across the organization to dispel such ambiguity of emergency alarms and indicators. Such practice would enable onboard personnel to respond appropriately during emergency and deal with such situations in a better way. The cost of developing uniformity in codes and practices is negligible compared to the damaging consequences it may cause by ignoring it.

8.2   OBJECTIVES

Alerts are of four types based on priorities, namely, (i) emergency alarm, (ii) alarm, (iii) warning, and (iv) caution, which are elaborated in Section 8.5. However, the scope of this study is limited to emergency alarm and indicators for offshore oil and gas installations.

The objectives of this study are the following:

(a) To develop uniform standards for alerts and indicators, especially emergency alarm system for offshore rigs, platforms, and installations across the organization

(b) To provide general guidance for selection or design of an appropriate emergency alarm system and indicators based on recognized international codes

As a corollary to the previous text, the study also aims at the following:

(c) To mitigate risk and eliminate safety threat to personnel onboard the oil and gas installation by dispelling confusion that arises due to different types of alerts and indicators prevailing on various offshore installations

(d) To minimize modifications for effecting uniform standard for emergency alarm system and indicators across the organization without compromising safety standards and ensuring least disturbances.

8.3 METHODOLOGY

The methodology followed in this study is described in the following paragraphs:

(a) The emergency alarm system of all offshore rigs, platforms, and major installations was studied. The code of signals for various situations was thoroughly examined.
(b) The available standards for emergency alarm system and associated code of signals at offshore oil and gas installations were extensively searched. API standard, US Federal standard/regulations, UK Offshore Installations standard/regulations, IEEE standard, OSHA standard, Norwegian petroleum industry standard, Canada Oil and Gas Production and Conservation Regulations, NFPA codes and standards, and Safety of Life at Sea (SOLAS) codes were referred to.
(c) HSE manuals of the company were studied, and discussions were held with HSE officials posted in the Assets, drilling services, and corporate office.

8.4 OBSERVATIONS

The "as-is" study of the existing system reveals the following:

(a) The emergency alarm system at different offshore rigs, platforms, and installations is not the same. The existing emergency alarm system and the corresponding code of signals at various offshore installations are shown in Appendix 8.A for drilling rigs and Appendix 8.B for production platforms.
(b) It may be seen from Appendices 8.A and 8.B that there are predominantly six emergency situations in drilling rigs, whereas platforms have few additional emergency situations. The code of signals for corresponding emergency situations varies widely at platforms and at drilling rigs. But it fares better at drilling rigs due to the reason explained at point "c".
(c) Drilling services had initiated the task of adopting uniform emergency alarm system on drilling rigs. But the approach followed by them was somewhat different—they had chosen one of the existing emergency alarm systems and codes of signal as model and replicated it at other installations. The emergency alarm system that were used at maximum number of installations was chosen as standard to minimize modification and change, replacement cost, and quicker implementation. But this approach has severe limitation and is questionable, which is discussed later in Section 8.5.

8.4.1 Reasons for Variations

The reasons for using different code of signals at various offshore installations are mainly due to the following:

(a) Different manufacturers and vendors had constructed and supplied the offshore rigs, platforms, and installations at different points of time in the past.

(b) The standard or exact specifications for emergency alarm system, indicators, and code of signals were not specified by the company to the manufacturers or contractors. As a result, different contractors had installed different code of signals and emergency alarm system as per their choice, convenience, and availability.

8.5 DEEP DIVE AND FINDINGS

As per International Maritime Organization (IMO) definitions (Section 8.3), alerts are basically announcement of abnormal situations and conditions requiring attention and are divided into four categories based on priorities, namely, (i) emergency alarm, (ii) alarm, (iii) warning, and (iv) caution. **Emergency alarm** indicates **immediate danger** to human life or to the ship and its machinery requiring immediate attention and action. **Alarm** is a high-priority alert requiring immediate attention and action to **maintain** navigation and operation of ship. **Warning** requires no immediate action and is presented for precautionary reasons. **Caution** is a low-priority alert for awareness of ordinary consideration [1]:

(a) Standard international code of signals for alerts (emergency alarm) and indicators was thoroughly surveyed as mentioned in Section 8.3b. It was observed that emergency situations are mostly industry specific and depend on types of operations. It differs from one type of industry to another. Individually, standards specified in API standard, IEEE standard, OSHA standard, UK Offshore Installations standard, US Federal standard, Canada Oil and Gas Production and Conservation Regulations, NFPA codes and standards, and SOLAS are not adequate for and have limited applicability to offshore oil and gas installations.

Different regulatory bodies have defined different sets of emergency situations depending on its activities and operations. For example, emergency situations at offshore drilling rigs and platforms are different from that of ships or seaborne vessels. Further, in case of ships, it may vary for passenger ship, ship carrying ordinary goods, dangerous chemicals carriers, nuclear merchant ships, high-speed craft, and so on. However, some commonality may be found among situations in various industries, for example, fire alarm and gas detection alarm.

(b) *No consensus on international code of signals for emergency alarm system and indicators could be found that covers a wide range of situations applicable for oil and gas installations.* However, mobile offshore drilling unit (MODU) code though limited in purpose and application is a useful guide.

IMO's "Code for the Construction and Equipment of Mobile Offshore Drilling Units (MODU), 2009 Resolution A.1023(26)" read with IMO's "Code on Alerts and Indicators, 2009 Resolution A.1021(26)" is a useful reference for designing and developing emergency alarm system for offshore oil and gas installations. MODU code is applicable to construction and equipment of MODUs and ensures safety of these units and personnel onboard, satisfying the requirement of 1974 SOLAS Convention.

Section 5.7.2 on "Alarms and Internal Communication" contained in MODU code, "Construction and Equipment, 2009 Resolution A.1023(26)," specifies that *the signals for alarm system should be limited to five situations, namely, (i) general emergency, (ii) toxic gas (H₂S), (iii) combustible gas, (iv) fire alarm, and (v) abandon unit signals* [2]. The standard for "fire alarm" and "gas detection alarm" under emergency alarm system specified in MODU code may be useful reference for developing uniform code for emergency alarm.

(c) The **general** points on **Code on Alerts and Indicators**, as specified in 2009 MODU Code A.1021(26): **Section 4**, may be kept in mind while designing and developing alerts (emergency alarm) and indicators [1], especially the following:

   (i) Subsections 4.1–4.3: Specify the general requirements of alerts and indicators (e.g., clear, unambiguous, audible and visual means, and so on)

  (ii) Subsection 4.8: Emphasizes on continuous power supply to alert system and automatic changeover to a standby power supply in case of disruption in normal power supply

 (iii) Subsection 4.11: Impresses on the need of designing the alert and alarm systems on "fail-to-safety" principle

 (iv) Subsection 4.15: Emphasizes on using fire-resistant cables for fire and general emergency alarms, public address systems, and power sources in high fire-risk areas

  (v) Subsection 4.18: Impresses on minimizing number of alerts and indicators on the navigation bridge

(d) Furthermore, the codes on **Audible Presentation of Alerts and Calls**, as specified in 2009 MODU Code A.1021(26): **Section 5**, may be helpful for designing and developing alerts (emergency alarm system) [1] for offshore oil and gas installations, especially the following:

   (i) Subsection 5.1: Impresses on clear and distinctive alerts all over the requisite spaces and need for visual indicators in the noisy spaces

  (ii) Subsection 5.4: Emphasizes on installing multiple audible signals or call devices in large spaces to ensure uniform sound level and to avoid shock in the vicinity of source of sound

 (iii) Subsections 5.10 and 5.11: Specify the limit of audible signal (<120 dB(A)), sound pressure level (1/3 octave band), and signal frequency (200–2500 Hz)

(e) The points on **Visual Presentation of Indicators and Calls**, as specified in 2009 MODU code A.1021(26): **Section 6**, may be a useful guidance for designing and developing indicators [1], especially the following:

   (i) Subsection 6.2: Specifies duration for illumination (at least 50% of the cycle) and pulse frequency (0.5–1.5 Hz)

  (ii) Subsection 6.6: Specifies the color (amber) and pulse frequency (at least 4 Hz) of supplemental visual indicators on mobile offshore drilling units (MODUs)

(f) **Alarms and Internal Communication** codes contained in **Section 5.7** of "Code for the Construction and Equipment of Mobile Offshore Drilling Units, 2009" A.1023(26) dwells on the following issues [2], which may be useful for developing the emergency alarm system for offshore oil and gas installations:

- Location and number of signals to be used for emergency alarm system (Subsections 5.7.1 and 5.7.2)
- Location for installation of public address system (Subsections 5.7.3 and 5.7.4)
- Internal means of communication (Subsection 5.7.5)
- Visual indicators in high-noise areas (Subsection 5.7.6)

(g) The aforementioned codes may be used as a basis by E&P companies and regulatory bodies to develop their own standard, as per their need and suitability. For example, the Norwegian petroleum industry has developed NORSOK standards, which are primarily based on recognized international standards with suitable changes and adding provisions that satisfy the need of Norwegian petroleum sector.

In this context, it is mentioned that NORSOK Standard T-100: **Section 5**, which defines the requirements of Public Address and General Alarm Systems [3] on offshore installations, is a useful reference for developing standards. It specifies, among others, the following:

- General requirement (Section 5.1)
- Performance including start-up (Subsection 5.3.1)
- Speech signal conditions (Subsection 5.3.2)
- Alarm tone signals (Subsection 5.3.3)
- Attention signals (Subsection 5.3.4)
- Audio coverage with the system in normal operation (Subsection 5.3.7)
- Flashing lights (Subsection 5.3.8)

(h) The Offshore Installations (Prevention of Fire and Explosion and Emergency Response) Regulations 1995—also known as PFEER—is followed in the United Kingdom [4]. It essentially deals with the emergency responses including escape, evacuation, rescue, detection, communication, mitigation, and so on. **Regulation 11** on **Communication** provides guidance on the emergency warning arrangements including tones of acoustic signals and colors of illuminated signs for oil and gas installations. In this regard, the following regulations may be noted:

  (i) Regulation 11(1): Emphasizes on appropriate arrangements for audible and visual alarm system for reaching to all persons on the installation

  (ii) Regulation 11(2) (a): Specifies the illuminated sign under varying circumstances, namely, in case of warning of toxic gas (a red flashing sign) and in all other cases (a yellow flashing sign)

  Regulation 11(2) (b): Stipulates acoustic signals for various types of warning such as warning for evacuation (a continuous signal of variable frequency), warning of toxic gas (a continuous signal of constant frequency), and for all other cases (an intermittent signal of a constant frequency)

(i) The US Code of Federal Regulations (CFR), **Title 46—Shipping: Part 113** "**Communication and Alarm Systems and Equipment**," is also a useful reference [5], especially the following sections and subsections:

- Subpart 113.25—General Emergency Alarm System: refer to § 113.25-5 (c) Location of contact makers in MODUs
- § 113.25-6 Power supply
- § 113.25-9 Location of general emergency alarm signal
- § 113.25-10 Emergency red flashing lights
- § 113.25-12 Alarm signals: Specify the sound pressure level for the emergency alarm tone at various spaces
- Subpart 113.50—Public Address Systems, § 113.50-15 Loudspeakers

(j) In order to have a broader overview of general alarm system in similar type of industry or operations, one may further consult the following rules and regulations:

   (i) SOLAS Chapter III Regulations 6.4 (Onboard Communications and Alarm), Regulations 6.5 (Public Address System) read with Life Saving Appliances (LSA code), and Regulations 7.2 (General Alarm and PA System, Recognizability of Audible Alarms) in cargo and passenger ships [6, 7].

   (ii) OSHA standard no. 1910.165 on employee alarm system [8].

   (iii) National Fire Protection Association (NFPA) codes [9], especially the following:
   - **NFPA 72**: National Fire Alarm and Signaling Code covers the application, installation, location, performance, inspection, testing, and maintenance of fire alarm systems, supervising station alarm systems, public emergency alarm reporting systems, fire warning equipment, and emergency communications systems (ECS) and their components.
   - **NFPA 110**: Standard for Emergency and Standby Power Systems covers performance requirements for emergency and standby power systems providing an alternate source of electrical power to loads in buildings and facilities in the event of failure of primary power source.
   - **NFPA 170**: Standard for Fire Safety and Emergency Symbols presents symbols used for fire safety, emergency, and associated hazards.

(k) As mentioned in Section 8.4c, drilling services of the offshore Asset under reference had chosen one of the existing emergency alarm systems and codes of signal as model for use in other installations. *But this approach is ridden with serious shortcomings, as it does not ensure compliance of two key aspects of developing standards, namely:*

   (i) *Conformance to internationally recognized codes and standards as mentioned in Sections 8.5a–j*

   (ii) *The requirement that alerts and indicators should be clear, distinct, and unambiguous*

8.6 RECOMMENDATIONS

Based on the discussions in Section 8.5, the following recommendations are made:

(a) In accordance with 2009 MODU code A.1023(26): Section 5.7.2 and prevailing practices at many offshore installations of the company as well as in other E&P companies, the *emergency situations may be broadly classified into six categories as follows: (i) fire/blowout, (ii) H₂S emission, (iii) hydrocarbon leakage, (iv) abandon* *platform, (v) man overboard, and (vi) all clear.*

All possible emergency situations may be classified into six categories, which are found adequate to cover possible emergency situations at offshore installations.

(b) The corresponding **code of signals** for the aforementioned emergency situations may be chosen from the following illustrative array of code of signals—*intermittent ringing, two short intermittent rings, three short rings, three long rings, seven short rings followed by one long ring, intermittent wailing, continuous ringing, long and short soundings, warble, yelp, hail and pass the word "man overboard," and so on.*

(c) The general code for presentation, design principles, and power supply of **alerts and indicators** shall be as per 2009 MODU code A.1021(26): Section 4; US CFR 46—part 113: subpart 113.25-5(c), § 113.25-6, which are elucidated in previous Section 8.5. Specific details such as *number of devices to be installed, location of devices, number of signals, sound pressure level, signal frequency, start-up, speech signal conditions, alarm tones and signals, attention signals, illumination sign, pulse frequency, color, and so on* may be in accordance with 2009 MODU code A.1021(26) (Sections 5 and 6); 2009 MODU code A.1023(26) (Section 5.7); NORSOK T-100 (Section 5); PFEER 1995: Regulation 11(1), 11(2) (b); US CFR 46—part 113: subpart 113.25, § 113.25-9, § 113.25-10; § 113.25-12, subpart 113.50, § 113.50-15, which are elaborated in previous Section 8.5.

(d) All offshore-going personnel including visitors must get familiarized with alerts (emergency alarm system), indicators, and code of signals, so that they can act appropriately in case of emergency.

(e) There should be a nodal team or group to implement and follow up the aforementioned recommendations. Generally, it is the responsibility of corporate HSE group and respective Asset HSE teams to adopt and implement uniform standard for alerts and indicators for all offshore rigs and installations. This would address the safety risks to offshore-going people to a great extent.

8.7 CONCLUSIONS

Safety threats, however, trivial or seemingly unimportant, shall not be ignored, especially in risky and vulnerable offshore operations. It was observed that different emergency alarm systems and code of signals prevailing at various offshore rigs and installations in an E&P company create confusion among crew members, contractor's personnel, and others who frequent different installations for work. It is a safety threat to offshore-going personnel and needs attention for immediate redressal.

This study tries to address the aforementioned safety risk to offshore-bound personnel. It recommends adopting uniform standard for emergency alarm and indicators for offshore installations across the organization:

(a) The various internationally recognized safety codes and standards were studied, but no "worldwide" or unique standard could be found that covers wide range of situations in oil and gas installations. Emergency situations vary from one type of business or activity to another and depend on many factors. Similarly, safety codes and standards of different regulatory bodies are found to be of limited use in terms of applicability and clarity.

(b) In view of this, many E&P companies have developed their own safety codes incorporating appropriate changes as per their need and suitability. Nevertheless, the following references are found useful guidance for designing and developing standards for alerts (emergency alarm system), indicators, and code of signals: IMO's 2009 MODU code A.1023(26). 2009 MODU code A.1021(26). UK PFEER 1995, NORSOK Standard T-100, US CFR 46—113, subpart 113.25, and so on.

(c) In accordance with 2009 MODU code A.1023(26) and objective of this study (see Section 8.2d), the emergency situations have been broadly classified into six categories for the purpose of standardizing emergency alarm system for offshore oil and gas installations. These are categorized as fire/blowout, H_2S emission, hydrocarbon leakage, abandon platform, man overboard, and all clear. The corresponding **codes of signals for each of these situations** have been suggested.

(The detailed recommendations are provided in Section 8.6.)

Chapter 9 focuses on optimizing supply chain management (SCM) system of an enterprise, which is important for smooth functioning of costly E&P operation. It diagnoses deficiencies in the current system and suggests ways to improve key SCM functions including procurement, inventory management, and SCM support services, to mention a few.

REVIEW EXERCISES

8.1 What are the different types of alerts? How are they differentiated?

8.2 What are the reasons for not having uniform codes of emergency alarm across the organization in general?

8.3 What safety risk has been discussed in this study?

8.4 What is the difficulty in adopting any one standard?

8.5 As a safety specialist, what are the emergency situations you would recommend for offshore installations in your organization?

8.6 As an expert, you have been asked to develop a uniform standard for emergency alarm system and indicators for offshore installations. How would you proceed?

8.7 State the applicability of the following rules and regulations:

(a) SOLAS

(b) IMO

(c) NFPA

(d) PFEER

(e) NORSOK

(f) OHSAS

8.8 In your opinion, which team should be responsible for ensuring uniformity of alerts and indicators across the organization?

APPENDIX 8.A   CURRENT EMERGENCY ALARM SYSTEMS/
CODE OF SIGNALS AT VARIOUS OFFSHORE RIGS

Sl. No.	Situation	Rig 1	Rig 2	Rig 3	Rig 4	Rig 5	Rig 6
1	Fire/blowout	Intermittent ringing	Intermittent ringing	Intermittent ringing	Intermittent ringing	Intermittent ringing	Intermittent ringing
2	H_2S emission	Two short rings intermittently	Two short rings intermittently	Two short rings intermittently	Two short rings intermittently	Two short rings intermittently	Continuous siren (Hooter type)
3	Leakage of hydrocarbon	Intermittent wailing	—	—	Intermittent wailing	Intermittent wailing	Continuous siren (Hooter type)
4	Abandon platform	Continuous ringing	Continuous ringing	Continuous ringing	Continuous ringing	Continuous ringing	Continuous ringing
5	Man overboard	Hail and pass the word "man overboard"	Hail and pass the word "man overboard"	Hail and pass the word "man overboard"	Hail and pass the word "man overboard"	Hail and pass the word "man overboard"	Hail and pass the word "man overboard"
6	All clear	Three short rings	Three short rings	Three short rings	Three short rings	Three short rings	Three short rings

Rig 7	Rig 8	Rig 9	Rig 10	Rig 11	Rig 12
Intermittent ringing	Intermittent ringing	Intermittent ringing	↑	↑	Intermittent ringing
Intermittent wailing	Two short rings intermittently	Two short rings intermittently		Prolonged intermittent sinus tone followed by announcement	
Intermittent wailing	—	Two short rings intermittently			Long and short soundings of alarm bell
			Seven or more short rings followed by one long ring on the general alarm bell followed by announcement	↓	↓
Continuous ringing	Continuous ringing	Continuous ringing		Announcement on PA system	Continuous ringing
Hail and pass the word "man overboard"	Hail and pass the word "man overboard"	Three prolonged blasts on the alarm bell with announcement		Prolonged intermittent sinus tone followed by announcement	Three long rings
Three short rings	Three short rings	Three short rings	Announcement on PA system	Announcement on PA system	Three short rings

APPENDIX 8.B   CURRENT EMERGENCY ALARM SYSTEMS/ CODE OF SIGNALS AT VARIOUS OFFSHORE PLATFORMS

Sl. No.	Situation	Platform 1	Platform 2	Platform 3	Platform 4	Platform 5	Platform 6
1	Fire/blowout	Warble	Siren	Yelp	Siren[4]*	Siren (continuous sinusoidal tone)	Yelp
2	H_2S emission	—	—	—	—	Yelp (interrupted fast wailing tone)	—
3	Leakage of hydrocarbon	Steady	Yelp	Steady	Steady		Pulse
4	Abandon platform	Siren	Pulse	Siren	Siren[3]*	Pulse (high pitch)	Warble
5	Man overboard	Hail and pass the word "man overboard"	Paging			Hail and pass the word "man overboard"	Hail and pass the word "man overboard"
6	All clear	Pulse	Steady	Pulse	Stutter	Steady	Steady
	Other Situations						
7	Evacuate	Yelp	Warble				
8	Emergency shutdown			Warble	Siren[5]*	Warble	Siren
9	Sprinkler						

Note: Sirens are of different types, which are denoted by superscript*.

Platform 7	Platform 8	Platform 9	Platform 10	Platform 11	Platform 12	Platform 13
Siren	Warble	Siren	Intermittent wailing of siren (siren)	Warble	Warble	Warble
—	—	—	—	Gong (high H_2S)/slow whoop (low H_2S)	Gong (high H_2S)/slow whoop (low H_2S)	
Yelp	Yelp	Yelp	Two short intermittent rings (high–low)	Steady	Steady	Steady tone
Pulse	Pulse	Pulse	Intermittent ringing of alarm bell (pulse)	Siren	Continuous siren	Siren
Paging	Paging	Paging	—	—	—	
Steady	Steady	Steady	Continuous ringing of alarm bell (steady)	Pulse	Steady sound	Pulse
				Yelp	Yelp	Yelp
			Slow whoop			

REFERENCES AND USEFUL LINKS

[1] International Maritime Organization (IMO), 2009, Resolution A.1021(26), Code on Alerts and Indicators, available at: http://www.imo.org/blast/blastDataHelper.asp?data_id=29981&filename=A1021(26).pdf (accessed on March 3, 2016).

[2] IMO, 2009, Resolution A.1023(26), Code for the Construction and Equipment of Mobile Offshore Drilling Units, available at: https://www.dpc.mar.mil.br/sites/default/files/ssta/modu_code.pdf (accessed on March 3, 2016).

[3] Norwegian Oil Industry Association, February 2010, NORSOK Standard T-100 Telecom Subsystems, 4th edition, available at: http://www.standard.no/en/PDF/FileDownload/?redir=true&filetype=Pdf&item=443201&category=4 (accessed on March 3, 2016).

[4] British Standards Institution, Health and Safety Executive, 1997, The Offshore Installations (Prevention of Fire and Explosion, and Emergency Response) Regulations 1995, Approved Code of Practice and Guidance, L65, 2nd edition, available at: http://www.hse.gov.uk/pubns/priced/l65.pdf (accessed on March 3, 2016).

[5] US Government Publishing Office, Federal Register, 1982 (updated weekly), The US Code of Federal Regulations, Title 46—Shipping Part 113, Communication and Alarm Systems and Equipment, available at: http://cfr.regstoday.com/46cfr113.aspx; http://cfr.regstoday.com/46cfr113.aspx#46_CFR_113pSUBPART_113p25 (accessed on March 3, 2016).

[6] Theunissen Technical Trading BV, 2006, Methods of SOLAS GA-Signal Distribution On-board Vessels, Fire Detection and Alarm Systems, available at: http://www.tttbv.nl/DocumentHandler.ashx?file=images%2FProductgroup%2FDocuments%2FEltek%2FELTEK+General+Alarm.pdf (accessed on March 3, 2016).

[7] International Maritime Organization (IMO), Sub-committee on Ship Design and Equipment, 2009, SOLAS Regulation III/6.4.3 and … Section 8.2 of Resolution A.1021(26) on Code on Alerts and Indicators, available at: http://www.iacs.org.uk/document/public/Publications/Submissions_to_IMO/PDF/CONSIDERATION_OF_IACS_UNIFIED_INTERPRETATIONS_pdf1991.pdf (accessed on March 3, 2016).

[8] Occupational Safety and Health Administration (OSHA), Standard 1910.165(a): Scope and Application, Emergency Alarm System, available at: https://www.osha.gov/pls/oshaweb/owadisp.show_document?p_table=standards&p_id=9819; https://www.osha.gov/SLTC/oilgaswelldrilling/ (accessed on March 3, 2016).

[9] National Fire Protection Association. Codes and Standards, NFPA 72: National Fire Alarm and Signaling Code, 2013; NFPA 110: Standard for Emergency and Standby Power Systems, 2016; NFPA 170: Standard for Fire Safety and Emergency Symbols, 2015, available at: http://www.nfpa.org/codes-and-standards/document-information-pages (accessed on March 3, 2016).

9

OPTIMIZING SUPPLY CHAIN MANAGEMENT SYSTEM OF AN E&P COMPANY

9.1 INTRODUCTION

In previous chapters, we have dealt with the improvement in core E&P activities and safety aspects. In this chapter, we would deal with an important functional area, namely, supply chain management (SCM) system, which has considerable influence on efficient operations of an organization. Smooth functioning of E&P operations is largely dependent on an efficient SCM system; even the slightest disruptions in material supply and delay in services may often prove costly to E&P companies.

SCM is the integration of various activities within and beyond the boundary of an organization that includes sourcing of materials and services, transforming them into intermediate or final products, and delivering them to its intended users and customers through a distribution system in an efficient and cost-effective manner. Therefore, SCM covers a wide spectrum of activities, such as procuring materials and services, integrating with production/manufacturing and related operations, managing warehouse and inventory system, distribution and delivery of final products, managing logistics, linking with finance and accounting system, and other related activities. The current study, however, focuses on some key functions of SCM that merit priority attention. The major shortcomings and areas of concern were identified through quantitative and qualitative assessment of these supply chain functions and benchmarking them with comparable industry standards. Accordingly, the following supply chain groups and their functions, subfunctions, and activities were studied for possible improvement:

- *Procurement group*: Purchase and contract, material planning, order processing, order execution, tendering, categorization of materials, and procurement of spares, items, and so on.

Optimization and Business Improvement Studies in Upstream Oil and Gas Industry, First Edition.
Sanjib Chowdhury.
© 2016 John Wiley & Sons, Inc. Published 2016 by John Wiley & Sons, Inc.

- *Materials management group*: Inventory management, warehouse functions, stock and replenishment, material receipt, material issue and consumption, disposal, and so on.
- *SCM support group*: Demand forecasting, strategic procurement (sourcing of critical items), vendor relationship management (VRM), prequalification and selection, vendor performance evaluation, strategic partnership, and so on.

(The distinction between function, subfunction, and activity is explained in Section 9.2.4 and in Figure 9.1.)

9.1.1   Background

E&P activities require capital-intensive and high-technology equipment, tools, spares, consumables, and specialized services. These are procured from various vendors across the globe and involve varying degrees of complexities, uncertainties, and risks. In order to keep pace with the growing E&P activities, there is a need to streamline and improve key supply chain functions and activities in the organization. This exercise is aimed to realize this purpose.

This study was conducted for a major E&P company whose operations are spread over several Assets and Basins in different geographical locations. It has onshore and offshore operations that have varying degrees of logistics challenges for supply of materials and services and transportation of heavy equipment and tools in remote and difficult locations, such as hilly terrains, forests, marshy land, thickly populated habitats, and so on.

Before delving deep into the aforementioned study of SCM system, it would be helpful to know about the evolution of SCM and different facets of material planning, which are described in the following paragraphs.

9.2   THE EVOLUTION OF SCM

9.2.1   Material Requirement Planning

Material planning has evolved over the years in tandem with the growing complexities of business. The need for material planning arose in the early twentieth century with the advent of mass manufacturing industries for maintaining uninterrupted production to meet the delivery schedule. Material planning at that time was mostly limited to determining safety/buffer stock, reorder point/level, lot size, economic order quantity (EOQ), demand forecasting, and so on, which was also known as inventory control. Until 1960, the focus was on inventory control and cost control. With rising competition and complexities in business, the need for integrating production planning, scheduling, and inventory control was felt, so that business operation would not falter for want of materials. Thus, *material requirement planning (MRP) was developed in 1964, which in broader terms may be defined as a computer-based information management system for integrated production planning, scheduling, and inventory control*. It uses master production schedule (MPS), bill of materials (BOMs), and inventory records for the following purposes:

(i) Identifying materials required to be ordered or replenished
(ii) Determining the order quantity, periodicity, and schedule of (placing) order for dependent demand materials (e.g., demand for components for manufacturing/assembly line)

These, in effect, ensure the following important tasks:

- Timely availability of material, items, and components required for continuous production.
- Delivery of intermediate/finished products to its intended users/customers.
- Maintain optimum inventory of materials to avoid stock-out situation and building up excess stock of materials (idle resources).
- Suggest production schedule, purchasing schedule, and other customized reports.

MRP gained prominence in 1970s and became a useful system in manufacturing and other industries. But it also had some of the following limitations:

- It does not take into account the capacity constraints.
- It is mainly concerned with coordination, planning, and procurement of raw materials for manufacturing, but does not consider financial and other aspects.
- There are data integrity issues concerning BOM, cycle count, scrap reporting, shipping and receiving errors, production reporting, replenishment, and so on.

In fact, the computer technology at that era had limited capability and was not advanced enough to handle these complex issues [2].

9.2.2 Manufacturing/Management Resource Planning

With the advancement of computing power, manufacturing/management resource planning called MRP II was developed in early 1980s to overcome the aforementioned limitations. MRP II is associated with integrating various aspects of manufacturing process, such as material planning, operational planning, financial planning, marketing, and so on. It is a detailed production schedule that can effectively:

(i) Handle fluctuation in forecasting data and corresponding readjustment in production planning

(ii) Consider capacity constraints of machines, labor, and related inputs for operational planning

MRP II is, in fact, modular-type construction and software-based applications, which are linked with a central database. It can simulate manufacturing system, generate various scenarios, and can answer "what-if" questions. It was a popular materials management system in the 1980s. But the computing power—hardware, software, and relational database technology—in the 1980s was neither powerful enough to provide speedy results in real time nor cost effective for business.

9.2.3 Enterprise Resource Planning

With the rapid advancement of information technology and computing power in the mid-1990s, enterprise resource planning (ERP) gained prominence. ERP is essentially a business integration information system in real-time operations and is the extension of MRP and MRP II linking the customers, suppliers, and other business entities. Both MRP and MRP II are still used independently in many industries and as embedded modules in ERP system. ERP integrates application programs in manufacturing, finance and accounting, sales and marketing, inventory and warehouse, engineering, purchase, suppliers, customers, and others through a common database. ERP is considered as successor of MRP and MRP II.

ERP solutions can be customized to meet specific business requirement of an organization. It helps in standardizing processes; encourages adoption of best practices; improves quality, accuracy, and speed of information; improves system integration in the organization; and reduces transaction cost. It also encourages communication and foster collaboration within and outside the organization. However, ERP packages are expensive and require great deal of efforts to customize to meet business requirement. Users (employees) need to be trained extensively for supporting implementation and subsequent operation. Generally, mid- and small-sized companies can not afford these; therefore, ERP is not suitable for small and medium enterprises.

9.2.4 Supply Chain Management

As mentioned in the previous paragraph, the focus until 1970 was on traditional mass manufacturing and related material planning system. With the rise in cost of production/operations, the focus shifted to cost optimization and inventory management in the late 1970s and 1980s. The emphasis during this period was on optimization of cost of materials and improving efficiencies in operations, logistics, and processes. This, in effect, led to development of various models for production and operations control, inventory management, and so on. All these evolving concepts and models focused on different segments of enterprise offering piecemeal solutions. The concepts of just-in-time (JIT) manufacturing/shipping, total quality management (TQM), and business process reengineering (BPR) gained prominence in the 1980s [1].

The concept of SCM emerged in the mid-1990s, which primarily aimed to integrate various resource plans of the entire enterprise. SCM integrates suppliers, manufacturers, distributors, logistics, retailers, and customers in an efficient and cost-effective manner. There are numerous definitions of SCM described by various authors and groups, such as the Institute for Supply Management, the Supply Chain Council (United States), Stanford Supply Chain Forum, MIT Center for Transportation and Logistics, and Simchi-Levi et al. [1] (see Bibliography). The essence of these can be summarized as follows.

SCM is essentially an all-encompassing approach of managing supply and demand; sourcing of materials, spares, and services; warehousing and inventory management; order execution; order management; distribution; and delivery to the desired destination leading to overall system optimization. Thus, SCM is the design and management of processes across organizational boundaries that aim to match demand and supply, maximize value to customers, and optimize total system cost.

SCM system comprises many functions, subfunctions, and activities. A **function** is a high-level event (e.g., purchase and contract), which can be further divided into **subfunctions** such as material planning, order processing, order execution and delivery, and so on. A subfunction, in turn, consists of various **activities** such as material categorization, procurement of spares, procurement of consumables and other items, tendering, bid evaluation, and so on. Therefore, functions, subfunctions, and activities differ in hierarchical and aggregation levels (refer to Fig. 9.1). The importance of functions, subfunctions, and activities varies from one type of industry to another. For example, procurement, distribution, product assortment, and pricing may assume more importance in retail industry/business than in manufacturing or oil and gas industry. Some of the typical functions, subfunctions, and activities associated with SCM are mentioned as follows:

- *Sourcing/procurement of materials and services*: Purchase and contract, material planning, order processing, order execution, and so on, as mentioned in Section 9.1
- *VRM*: Vendor selection, vendor performance evaluation, strategic partnership, and so on
- *Manufacturing*: Production planning, scheduling, warehouse management, inventory control, and so on
- *Distribution system*: Trucking (mostly used), railroads (for large loads), waterways (for bulky, low-value cargo), airfreights (for light loads), pipelines (for oil, gas, chemicals), and so on
- *Logistics*: Transport operators/contractors and others
- *Finance and accounts*: Credit and cash transfers, accounts payable and receivable, and so on

The relevance and importance of various supply chain functions, subfunctions, and activities as mentioned previously depend on the industry type, size of the company, transaction volume, operational and environmental requirement, and so on.

To learn more on SCM, material planning, and related issues, interested readers may refer to "Bibliography" presented at the end of this chapter.

9.3 OBJECTIVES

The competitive business world nowadays requires an efficient SCM system to suit need and objectives of the organization. Material planning and procurement process in the E&P enterprise under study is found to have comparatively long procurement cycle time, high inventory level, and slow decision-making process for procurement and has extensive emphasis on procedures and documentations. In view of the aforementioned, the specific objectives of this study are as follows:

- To assess the maturity level of key supply chain functions in the organization and identify potential gaps
- To streamline key supply chain functions to match pace with the growing E&P activities
- To rationalize procurement process and reduce procurement cycle time by streamlining activities and process
- To optimize inventory and improve warehouse functions and activities
- To rationalize SCM support services for better performance, efficiency, and focus

The auxiliary objectives of this study are as follows:

- To identify the areas of concern and find out their underlying causes
- To analyze material consumption pattern and inventory level and suggest ways of cost optimization
- To study the existing SCM setup in the organization and suggest measures for improvement

9.3.1 Methodology

The methodology followed in this study is described in the following paragraphs:

(a) The maturity of various supply chain functions was assessed using quantitative methods for analyzing historic and real-time data and processes.
  - Inventory data for 10 years (say, Year N to Year $N-10$) were captured and analyzed.
  - Material procurement and consumption data for 2 years (say, Year N and Year $N-1$) of offshore operations of the company covering three important Assets, one Basin, and services were studied and analyzed.
  - Most of the data is sourced from the existing ERP system, concerned unit/ department, published annual report of the company, and so on.
(b) The collected data and information was supplemented with qualitative assessment by interviewing and discussing with end users and key SCM officials in the organization to evaluate the maturity of the current supply chain functions.
(c) The results were benchmarked against comparable industry standards and best practices to identify gaps and potential areas of improvement.
  - Performance metrics of SCM functions were computed and compared with that of other national oil companies (NOCs), international oil companies (IOCs), upstream sector, and best practices.

9.4 SCM SYSTEM

The outline of the SCM system studied in this work is shown in Figure 9.1. It contains important SCM functions, subfunctions, and activities, which are discussed in subsequent sections. SCM system in the organization under reference may be divided into three main groups, namely, (i) procurement group, (ii) materials management group, and (iii) SCM support group. Each group consists of number of functions, for example, procurement (purchase and contract), materials management (inventory management and warehouse), and SCM support (demand forecasting, VRM, and strategic sourcing). Each function, in turn, contains a number of subfunctions and activities as illustrated in Figure 9.1.

Before we proceed further, it is important to know material classification, which differs from company to company. Different companies classify it in a different way for the purpose of procurement and inventory management. For the organization under

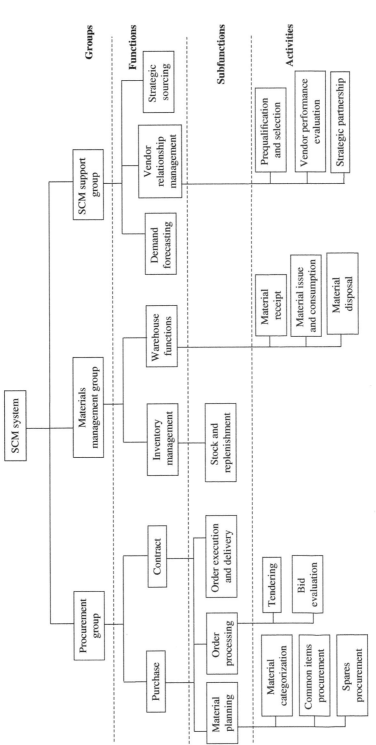

FIGURE 9.1 SCM system (functions, subfunctions, and activities).

reference, it is broadly classified into three categories, namely, **stores**, **spares**, and **capital items**. There are many subgroups under each of this group. The typical stores items are drill pipes, casing pipes, drill bits, and other drilling items; electrical materials, cables, fittings, and others; well heads and Xmas trees; and chemicals, oil well cement, POL, and others. **Chemicals** referred to in this study is a subgroup of stores items.

Apart from these, items costing a certain amount (say, USD 100) or more and with a life of more than a certain period (say, 1 year) are categorized as **capital items** for the purpose of procurement and accounting. "Stores" and "spares" are considered having a life of less than 1 year.

The aforementioned groups, functions, subfunctions, and activities are discussed in the following sections: Section 9.5 (procurement), Section 9.6 (materials management), and Section 9.7 (SCM support).

9.5   PROCUREMENT GROUP

There are three distinct phases in procurement cycle of materials and services, namely, (i) requisitioning, (ii) ordering, and (iii) delivery. There is considerable scope for improvement in the process of "requisitioning" and "ordering," which are internal to the system. But "delivery" is external to the system and can be improved to some extent through fostering strong relationship with suppliers.

9.5.1   Procurement Cycle Time

Procurement cycle time consists of three main components, namely, material planning process, ordering process, and delivery times. It is the sum of these three components, that is, procurement cycle time = material planning process time + order processing time + delivery time.

The procurement cycle times for different groups of items are found to vary between **11** and **17 months** with a mean time of **15 months**, which are detailed as follows: stores items—**14–16 months**; spares—**13–15 months**; chemicals—**11–13 months**; and capital items—**15–17 months**.

This is in conformity with the observation that the average time taken for material planning process is 4 months (refer to Section 9.5.2), the average order processing time is 5¼–8¾ months for different groups of items with a mean time of approximately 7 months (refer to Section 9.5.3), and the average delivery time of materials is 3–4 months.

Thus, average procurement cycle time = 4 months (material planning process time) + 7 months (order processing time) + 4 months (delivery time) = 15 months.

These are much higher than that of similar national E&P companies (3–10 months) and regional E&P companies (0.7–4 months, with an average of 2.2 months), which are shown in Figure 9.2.

This section discusses various components of procurement cycle including key functions, subfunctions, and activities such as material planning process, order processing, tendering, procurement of spares and common items, categorization of materials, and so on.

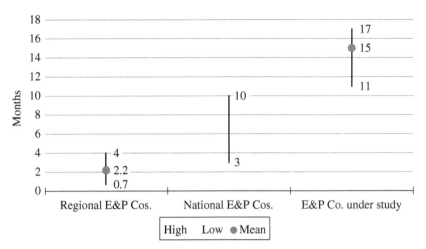

FIGURE 9.2 Procurement cycle time (months—high, low, and mean).

9.5.2 Material Planning Process

The present practice of material planning process ("as-is" processes) is described in the following paragraphs:

(i) Annual requirement of all items, namely, capital, spares, stores, and consumables, are invited from each unit/work center.

(ii) This is consolidated at base by procurement and provisioning cell of each Asset/Basin/service.

(iii) Net requirement is arrived at after validating the available stock, quantity on order and in pipeline.

(iv) Proposal as well as purchase requisition (PR) is initiated by provisioning cell of respective Asset/Basin/services for approval.

(v) Requisition process is delayed, if the last purchase price (LPP) is not available for some items. In such cases, budgetary quotations are invited from at least three parties.

(vi) Financial concurrence is obtained.

(vii) Administrative approval and sanction of the competent authority are obtained subsequently.

(viii) It is released by MRP controller, followed by finance and approving authority.

(ix) PR is released by the indentor.

(x) Purchase order(ing) procedure begins.

The existing material planning process is shown in Figure 9.3. It takes around 4 months to complete this process. At times, the process of requisition and approval become cumbersome.

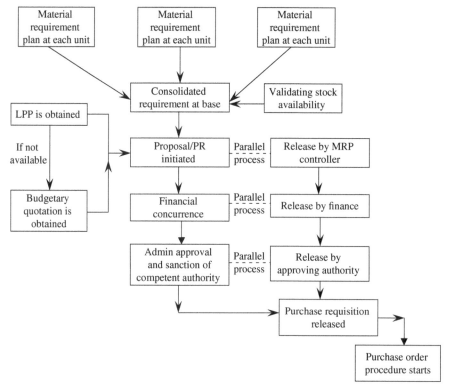

FIGURE 9.3   Present ("as-is") material planning practices.

9.5.2.1 Deficiencies in Material Planning Process: The deficiencies identified in material planning process are summarized as follows:

- (i) Consumption norm for regularly consumed items is not clearly defined.
- (ii) Justifying requirement of each item at each business unit/work center takes a long time.
- (iii) The requisition process is *mostly manual and at times person specific.*
- (iv) Paper requisition at business unit/work centers and consolidation at base consume considerable time.
- (v) The process is repeated both during budget preparation and case-by-case approval stage.
- (vi) Due to long procurement cycle time, requirement is often inflated by the indentor.
- (vii) Repeated tendering process takes place even for regularly consumed items.

9.5.2.2 Suggestions: The following suggestions are offered to improve the procurement and material planning process:

- (a) Many functionalities of ERP system are not being fully utilized currently, which may be gainfully utilized. *Manual material planning process and demand forecasting based on paper requisition, trailing, and consolidation, which is time consuming and often found inflated, may be dispensed with. This may be replaced with system-generated annual consumption standard for regularly consumed stores and spares items whose consumption can be predicted* with a fair degree of accuracy. For example, there is provision of MRP process in the ERP system for generating future demand of materials based on past consumption data. This will reduce manpower requirement for provisioning cell of each Asset/Basin/service and save considerable procurement cycle time (about 3–4 months).

(b) Material planning process may be integrated with the annual budgeting process, so that once budget is approved, the practice of case-by-case approval for material procurement may be dispensed with. However, this would require more rigorous preparation of annual budget.

(c) The users' group may raise indent with the approval of the team/department head for procurement of materials within the allocated budget, without routing through the finance team/department and obtaining sanction again. This would save considerable time and make the process less cumbersome. Periodical statement may be sent to the finance group for updating utilization of budget.

(d) For uniform and speedy budget/approval process, it may be ensured that each proposal contains relevant information. *A standard checklist or standard format may be introduced for this purpose.* The checklist must contain the following:

   (i) Consolidated requirements taking into account stock position and material in pipeline

   (ii) Consumption pattern of items for the last 3 years and future projections

   (iii) Rate estimation and its basis (LPP/budgetary quote (BQ)/consultant)

   (iv) Budget provisions

(e) The use of email as a mode of communication may be encouraged for obtaining BQ, technical clarification, and so on. This would reduce the processing time. Industry database may also be used for cost estimate and budgeting.

9.5.3 Order Processing Time

Order processing time, which is the time between PR and purchase order (PO) in the organization, is found to vary between **165** and **269 days** with an average time of **205 days**. The norm for order processing time fixed by the company is between **75** and **115 days**. These are schematically shown in Figure 9.4.

The average order processing time for **stores, spares, chemicals**, and **capital** items in the **organization as a whole** is found to be **244, 207, 157**, and **261 days**, respectively. The same for stores, spares, chemicals, and capital items for **offshore Assets** under study is observed: **327, 215, 226**, and **239 days**, respectively.

It may be noted from the previous text that the actual order processing time in the company (as well as offshore Assets) is way above the norm time, which is shown in Figure 9.5.

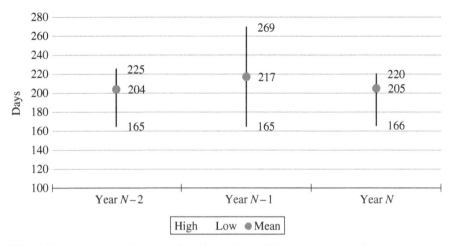

FIGURE 9.4 Average order processing time (days—high, low, and mean) (current company norm time: 75–115 days).

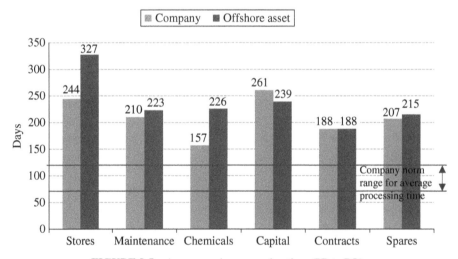

FIGURE 9.5 Average order processing time (PR to PO).

9.5.3.1 Tendering

9.5.3.1.1 Observations: The observations related to tendering process and activities are presented as follows:

(a) About 460 tenders are floated annually for procurement of around 900 items in offshore Assets and services. Out of these, around 5–10% items are procured through retendering every year. The reasons for retendering are as follows:
  • Lack of competition
  • Technically not acceptable bids (mostly found in nonoperational items such as tube lights, bulbs, etc.)
  • High price
  • Representation from aggrieved bidders in some cases

In order to reduce or eliminate retendering, it is necessary to encourage participation of qualified/empaneled bidders and enhance competition.

(b) *Limited Response from Qualified Bidders*

There were limited responses from qualified suppliers in offshore Assets during Year N and Year $N-1$, which are elaborated in the following text:

- There was only one qualified bidder for procurement of 16 items valuing USD 4 million.
- There were two qualified bidders for procurement of 41 items costing USD 7 million.
- There were three qualified bidders for procurement of 20 items costing USD 2.5 million.

These are depicted in Figure 9.6.

(c) *Reasons for Limited Response*

The reasons for limited response are mentioned in the following text:

(i) Low volume and/or price of order.

(ii) Rigid tender terms and conditions, technical specifications, and voluminous tender document.

(iii) It is observed that for some items like flexible steel hose, coil tubing reel, and so on, only one or two bidders had responded in the last few years.

(d) *Bid Evaluation*

Being a state-owned E&P company, the organization is guided by government rules, regulations, and procedures for procurement of materials and services. There is little scope for deviating from these, and it may not always result in best commercial deal for procurement. But the government procedures ensure that the company has a transparent process of approval and follows it for selection of suppliers.

The organization mostly follows two bid systems, namely, (i) technical bid and (ii) commercial bid for high-value tenders. Only technically qualified bids are considered for evaluating the commercial offer, and the lowest priced bid is generally selected for award of the contract. But the lowest priced bid is often found, as not the best commercial offer. By virtue of its statute, it is difficult for the company to deviate from the prevailing government rules, regulations, and procedures.

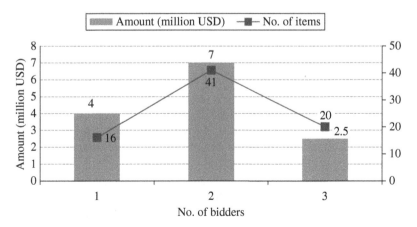

FIGURE 9.6 Number of bidders versus number of items.

9.5.3.1.2 Suggestions: In order to overcome the shortcomings of tendering process and related activities, the following recommendations may be followed:

(a) In order to avoid repeated tendering process, more number of items of regular consumption should be identified and brought under rate contract.

(b) Wherever feasible, tenders may be invited for the supply of materials for a longer period (say, 3 years or so) with staggered delivery, especially in those cases where market has stabilized.

(c) *Limited tenders may be floated for procurement of regularly consumed items among empaneled vendors. The clauses like bid bond, tender fee, performance guarantee, and so on, may be waived for empaneled suppliers.*

(d) The reason for lack of competition may be further analyzed. If it is due to stringent terms and conditions or experience criterion, the same may be relaxed to enhance competition without compromising quality and functional value.

(e) Apart from publishing in the electronic media/website, the copy of the tender document may also be forwarded to the prospective bidders/suppliers.

(f) All issues may be explained properly and exhaustively in prebid conferences, so that chances of rejection over trivial issues are minimized. Bidders may be requested to take the prebid conference seriously.

(g) Early and wide implementation of e-procurement may be emphasized.

VRM and Prequalification Work

(h) In order to reduce procurement cycle time, reliable sources of supply may be identified for various items based on predefined criteria. Open tenders may be invited for empanelment of reputed suppliers. Limited tenders may be floated among empaneled vendors for procurement of regularly consumed items.

(i) The empanelment of suppliers may be done centrally by SCM support group and reviewed every year. A centralized cell may be formed for this purpose.

(j) Suppliers' performance and rating may be introduced, so that good and qualified vendors get preference.

(More on VRM and prequalification work are discussed later in Section 9.7.4.)

Bid Management

(k) The present bid document may be simplified and made more concise, and documents required to be submitted by bidders may be reduced, wherever possible.

(l) A cross-functional team consisting of representatives of the procurement team/department and the legal and finance group may be constituted to review and simplify the bid document and tender conditions. Bid bond, tender fees, and liquidated damage (LD) clauses may be waived for low-value tender—valuing less than a specified amount (say, USD 70,000).

(m) The payment terms may be relaxed, and advance payment as well as easy-stage payment, wherever possible, may be allowed.

(n) The escalation of price due to change in critical inputs such as diesel and steel may be allowed by incorporating a suitable clause/formula.

Bid Evaluation

(o) Enforce rigorous and stringent technical qualification criteria, so that only reputed and highly capable companies participate and qualify.

(p) Invite tenders only from the empaneled/reputed and short-listed/approved companies.

9.5.3.2 Delay in Order Execution: The order execution is mostly found delayed. The reasons for delay in order execution are mainly due to the following:

(i) High emphasis on procedure (procedures are archaic)

(ii) Extensive documentation

(iii) To-and-fro file movement between indenting team/department and materials and finance group for justification, clarification, and so on

(iv) Numerous approvals from tender committee and others

(v) Delay in technical evaluation

(vi) Representation from aggrieved bidders

(vii) Poor response by original equipment manufacturers (OEMs) due to the following reasons:
- Extensive procedural requirement by the company
- Low quantity of orders in some cases
- Reputed OEMs/suppliers that are reluctant to comply with many conditionality in the tender such as tender fee, bid bond, performance bond, and so on

9.5.4 Procurement of Common Items

9.5.4.1 Observations: The following observations are made for the procurement of common items in offshore Asset 1, Asset 2, and Asset 3:

- About 7560 unique items were procured in Year N by three offshore Assets, namely, Asset 1 (A1), Asset 2 (A2), and Asset 3 (A3).
- 66 items common for all three offshore Assets were procured individually by these Assets.
- 310 items common for any two Assets were procured individually by these Assets. These are schematically shown in Figure 9.7.

9.5.4.2 Suggestion: The following suggestion may be followed to improve the procurement of common items in Assets/region:

(a) The common items of all three offshore Assets, services, and offices should be procured centrally by one of these Assets. Alternatively, a centralized procurement cell may be formed to cater the requirement of all three Assets in the region.

9.5.5 Procurement of Spares

9.5.5.1 Observations

Pattern of Spares Procurement: The analysis of procurement of spares in offshore Assets during Year N and Year $N+1$ reveals many interesting facts, which are mentioned in the following paragraphs.

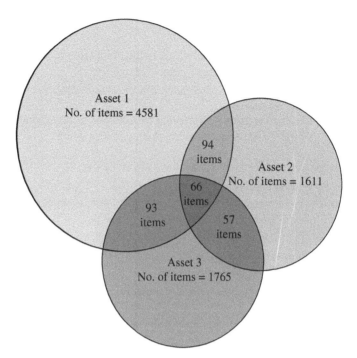

FIGURE 9.7 Procurement of common items.

10,728 pares items were procured during Year N:

- Only 1 unit of 3217 different items was procured, which accounts for 30% of spares procurement costing USD 19 million.
- Only 2 units of 2134 different items were procured, which accounts for 20% of spares procurement costing USD 12 million.
- Up to 6 units were procured for 7736 different items, which accounts for 72% of spares procurement costing USD 53 million.
- More than 10 units were procured for 2117 different items, which accounts for 20% of spares procurement costing USD 20 million.

These are graphically shown in Figures 9.8 and 9.9.
Consumption versus Procurement of Spares

(a) During Year N, spares around *USD 34 million* were consumed at offshore Assets/ Basins/services as against procurement of *USD 54.5 million*. The excess procurement was made in Asset 1 (*USD 14 million*) and drilling services (*USD 7 million*). The breakdown of procurement and consumption is shown in Figure 9.10.

(b) The low consumption of spares is due to the nonavailability of equipment for planned overhaul because of operational necessity.

(c) It also indicates tendency of overprovisioning or inflating requirement to be on the safe side.

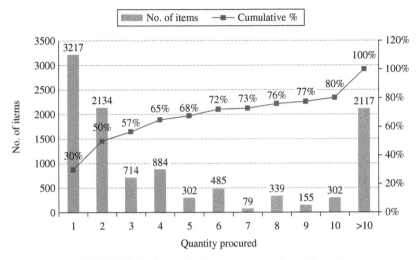

FIGURE 9.8 Analysis of spares procured (# of items).

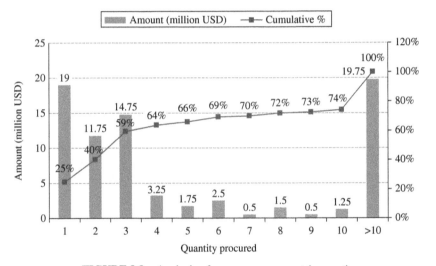

FIGURE 9.9 Analysis of spares procurement (amount).

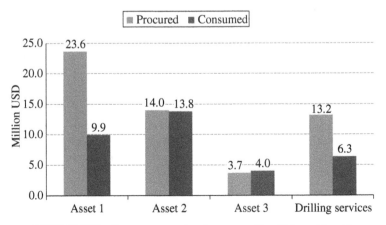

FIGURE 9.10 Spares procured and consumed in offshore, Year *N*.

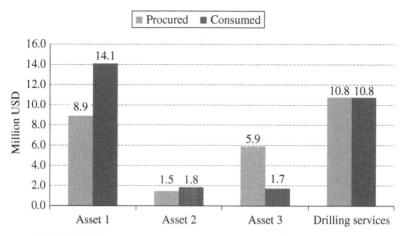

FIGURE 9.11   Spares procured and consumed in offshore, Year $N+1$.

(d) During Year $N+1$, spares consumption at offshore Assets/Basins/services was marginally higher (USD 28.5 million) than the procurement amount of USD 27 million. Further breakdown of procurement and consumption is shown in Figure 9.11.

Procurement from OEM

(e) There are nearly *348 OEMs* for mechanical, electrical, and instrumentation items in *Asset 1 alone*. Proposals are processed OEM wise for sanction and involve considerable administrative work and consume time. 198 cases having less than 4 items in each case were processed in Year N in Asset 1.

Requirement of such low volume and/or cost of spares in each case evokes poor response from many OEMs. At times, the prices quoted by them are high. All these result in delay in procurement and affect scheduled maintenance plan.

(f) During Year $N+1$, about 5031 unique items were procured through 443 POs by drilling services for all three Assets. Out of which, 681 items (13.5%) were procured through multiple POs, most of which are petty operational items.

(g) During Year $N+1$, about 3100 unique items were procured in Asset 1 through 271 POs. Out of which, 178 items (5.7%) were procured through multiple POs.

9.5.5.2 Suggestions: The following suggestions would help in rationalizing and improving procurement of spares, which is crucial for uninterrupted operations:

(a) Instead of processing proposal for each OEM or its authorized distributor and contacting them individually, a few authorized warehouses may be identified who could supply different types of spares (even if their consumption volume and/or cost is low).

Authorized warehouses must provide respective OEM's test certificate and warranty cover. This would eliminate processing of numerous cases and contacting different OEMs and would ensure timely availability of spares.

(b) Encourage rate contract and O&M contract with OEMs or authorized supply houses for supply of spares, technical services, overhauling of equipment/machineries, and so on, for a longer duration with agreed terms and conditions including provision for discount, escalation of prices, and so on.

(c) Authorized workshop (mechanical, electrical, instrumentation, and automobile) may be empaneled in each work center/Asset/Basin, as per need for work-up to a specified value (say, USD 100,000) with agreed rates, terms, and conditions.

(d) Priority may be accorded to plan overhauling of equipment.

(e) Dispose off unwanted/nonmoving spares (which is elaborated in Section 9.6.2.2.2).

9.5.6 Categorization of Materials: Core and Noncore Procurement

9.5.6.1 Observations: The following observations are made on categorization of materials and procurement of core and noncore items:

(a) There is no difference in the process, priority, or focus for procurement of materials and services of core and noncore businesses in the company. The core business accounts for over 90% of annual spend, but the number of contracts is nearly same for both core and noncore businesses. Therefore, much time is spent for the procurement processes of noncore businesses, which otherwise could have been focused and better utilized for core businesses.

(b) It is observed that the procurement of critical and high-value drilling items account for nearly three-fourths of annual spend, which are procured through less than 5% of POs issued in a year. On the contrary, noncritical, low-value fast-moving items that account less than 5% of annual spend are procured through approximately 50% of POs issued in a year.

(c) The number of POs issued for noncritical items is considerably high and consumes much time and resources of the organization.

(d) Multiple POs are issued to the same suppliers with irregular periodicity, and there is ample scope for consolidation.

9.5.6.2 Suggestions: The following suggestions are offered to improve categorization of materials and separate procurement of core and noncore items:

(a) In order to have better focus and priority for core businesses, the procurement of core and noncore businesses may be separated into two or more teams depending on the workload, volume of transactions, and size of Assets/Basins/region. Items may be categorized based on their functionality, criticality, and so on.

(b) Noncritical items with low value and short lead time may be procured through long-term contract/blanket PO, which would ease the burden of administrative work and allow the existing manpower to focus on procurement of critical and high-value items.

(c) Consolidate POs for the same vendors, and develop and maintain a schedule of PO release to top vendors. These will reduce the number of POs, administrative work, and procurement cycle time.

9.6   MATERIALS MANAGEMENT GROUP

As mentioned earlier, this section discusses inventory management and warehouse functions including stock maintenance and replenishment, material receipts and consumption, disposal of surplus materials, and so on.

9.6.1   Inventory Management

9.6.1.1   Observations and Areas of Concern: The following observations are made on the inventory holding and warehouse functions of the company:

(a) *Inventory (as % of Capital Employed)*

Inventory holding of the company for the past 11 years is shown in Figure 9.12. It may be seen that inventory holding increased from 4.4% of capital employed in Year $N-3$ to 6.1% in Year N. This is high compared to inventory holding of similar national E&P companies (0.5–2.4%) and regional E&P companies (average 3.1%) (Fig. 9.13).

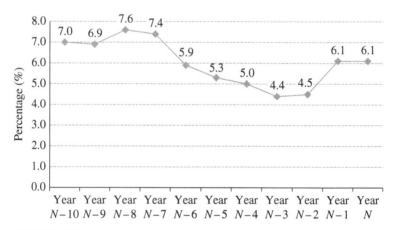

FIGURE 9.12   Inventory as percentage of capital employed in the company.

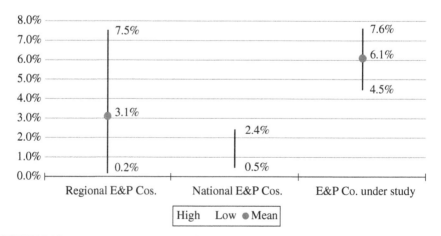

FIGURE 9.13   Inventory as percentage of capital employed: comparison (high, low, and mean).

(b) *Inventory (in Terms of Stock-Month)*

Inventory in terms of stock-month in the company has increased from 9.3 in Year $N-3$ to 15.8 stock-months in Year N (Fig. 9.14).

(c) *Inventory Turnover*

Inventory turnover of the enterprise for the last 11 years is shown in Figure 9.15. The annual inventory turnover rates (i.e., revenue/inventory) in the company is low and ranges between 6.8 and 22.5. It has gone up lately due to rise in oil price.

(d) *Reasons for Rise in Inventory*

The following reasons are responsible for the rise in inventory in the recent years (during Year $N-3$ and Year N):

  (i) The ERP system is under the process of implementation. Prior to the implementation of ERP system in the organization, materials issued from stores/warehouses were considered as consumed, even though the materials

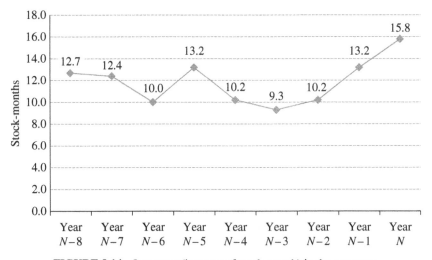

FIGURE 9.14   Inventory (in terms of stock-month) in the company.

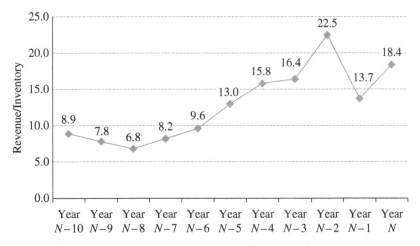

FIGURE 9.15   Inventory turnover (revenue/inventory) in the company.

were kept at different work units/locations for anticipated work. But after introduction of ERP system in the recent years, these unused items have been captured as inventory at locations, resulting in the rise in inventory.

(ii) Each Asset/Basin is procuring most items including common items separately and maintaining separate inventory.

(iii) The accumulation of unwanted spares is due to obsolescence and replacement of equipment or unserviceable condition of the equipment, for example, spares of DCS, EBARA pump, paging system, and so on.

(e) *Inventory Distribution Pattern*

There are total 18,041 inventory items in three offshore Assets, namely, Asset 1, Asset 2, and Asset 3. Of these, spares account for 90.4% (16,301 numbers), followed by stores items 8.9% (1614 nos.) and capital items 0.7% (126 nos.).

These are shown in Figure 9.16a (offshore as a whole) and for individual Assets (Fig. 9.16b–d). The distribution pattern follows the same trend for individual Asset as well as offshore as a whole.

(f) *Valuation of Inventory (in Offshore Assets)*

The analysis of inventory valuation at offshore Assets reveals the following:
• Stores items account for 57% (USD 50.6 million) of inventory value, even though its share is 8.9% of the total number of items.

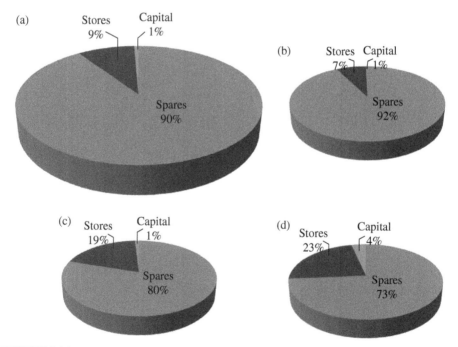

FIGURE 9.16   (a) Inventory: category-wise distribution (no. of items), (b) Asset 1, (c) Asset 2, and (d) Asset 3.

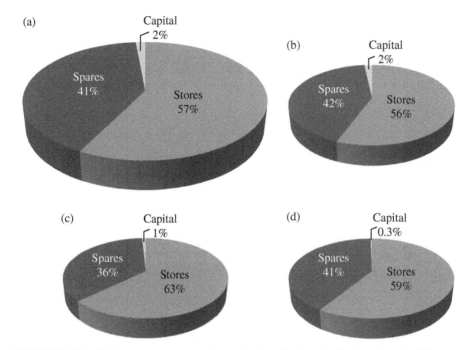

FIGURE 9.17 (a) Inventory: category-wise valuation, (b) Asset 1, (c) Asset 2, and (d) Asset 3.

- The spares account for 41% (USD 36.1 million) of inventory value and consist 90% of the total number of inventory items.
- The value of capital items is only 2% (USD 1.34 million) of inventory and comprises merely 0.7% of the total number of items.

  These are shown in Figure 9.17a (total offshore) and individual Assets (Fig. 9.17b–d). The same pattern is prevalent in other Assets and Basins across the company.

9.6.1.2 Suggestions: The following suggestion would help in optimizing inventory:

(a) The users' group/indentor ought to be the owner of inventory and would be responsible for excess buildup of stock. Materials group would be responsible for the inventory in transit. The users' group/indentors will be answerable if the stock of an item exceeds "maximum" level.

9.6.1.3 Stock Maintenance and Replenishment
9.6.1.3.1 Observations: The inventory segmentation practice is followed in the company, for example, ABC analysis of materials as per value, moving type (slow and fast), requirement (vital, essential, and desirable), and so on. But these are not gainfully utilized for streamlining inventory stock and service level. Traditional requisition process is in vogue for replenishment orders, and the advantage of system-generated replenishment level/order is not availed.

Furthermore, the integration between materials management and "maintenance and operations" planning is predominantly a manual, tedious, and time-consuming process.

9.6.1.3.2 Suggestions: The following suggestions are made to address the previously cited issues:

(a) Introduce system-generated replenishment level/order and safety stock with validation from user group taking into account future work plan. The various inventory control parameters like safety stock, reorder level, EOQ, and so on, may be determined based on the past consumption data available in the ERP system. These may be compared with the existing safety stock, reorder level, and so on, and reviewed accordingly.

(b) Automate system integration between materials management and "maintenance and operations" planning for faster response, shorter lead time and optimum inventory. The existing ERP system may be used for this purpose.

9.6.2 Warehouse Functions

Warehouse functions studied in this work include material receipt, material issue and consumption, and disposal of surplus materials, which are discussed in the following subsection.

9.6.2.1 Material Receipt and Consumption Pattern

9.6.2.1.1 Observations: Material receipt and consumption pattern reveal the following:

(a) It is observed that receipt as well as consumption of materials increase considerably (about 40–100%) during the last quarter of the fiscal year compared to the remaining quarters. The rise in receipt of materials in the last quarter is mainly due to the realization of materials in transit (MIT) during the year-end. The quarterly material receipt and consumption pattern in total offshore are shown in Figure 9.18a, and that of its constituent Asset 1 and Drilling Services are shown in Figures 9.18b and c, respectively.

(b) The consumption of materials increases considerably in the last quarter of the fiscal/budget year. The high material consumption during the last quarter is mainly due to the fact that on many occasions materials consumed during operation/ activity are not immediately posted; these are reconciled during the year-end.

(c) The consumption of materials is closely linked with the volume of work and activity. The level of activity remains nearly same in each quarter. Therefore, the consumption of materials should ideally be even in each quarter. Moreover, it is found that materials released from warehouses are nearly same in each quarter.

9.6.2.1.2 Suggestions: The following suggestions would help in addressing the aforementioned issues:

(a) Active persuasion and all-out efforts shall be made for higher liquidation of MIT round the year by the concerned department.

(b) The consumption of materials should be posted by indentors as soon as it is consumed or operation takes place. This will also help in material planning and drawing delivery schedule.

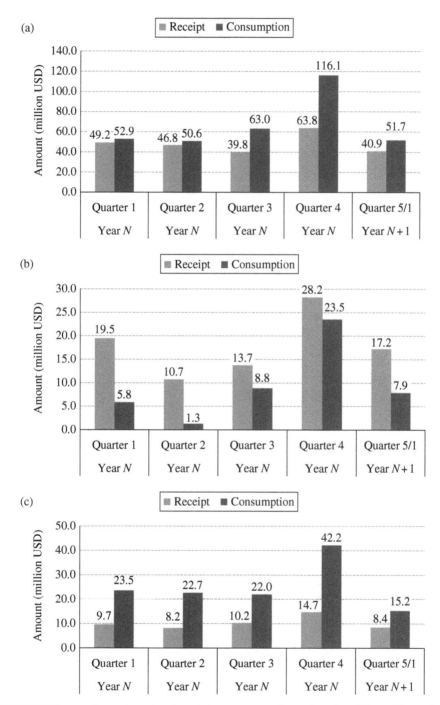

FIGURE 9.18 (a) Quarterly material receipt and consumption pattern in offshore, (b) quarterly material receipt and consumption pattern in Asset 1, and (c) quarterly material receipt and consumption pattern in Drilling Services.

9.6.2.2 Disposal of Surplus Materials

9.6.2.2.1 Observations: The practice of checking stock and declaring surplus materials is followed in all Assets/Basins. But it is done intermittently, and there is a lot of scope for improvement in terms of periodic and regular declaration, acceptability by other Assets/Basins, and disposal of these items.

There are few takers for most of these items. This is because either the items are obsolete or other Assets/Basins have already placed orders or procured these items.

9.6.2.2.2 Suggestions: The following suggestion may be useful to improve the disposal of surplus materials:

(a) The list of surplus materials of each Asset/Basin may be compiled and circulated across the organization. The excess materials may be transferred based on the need of respective Assets/Basins. The rest may be disposed off following the company rules. The chances of disposal of surplus materials and acceptance by other Assets/ Basins would improve, if such list is circulated regularly at a fixed interval (say, annually or biannually).

9.7   SCM SUPPORT GROUP

This section discusses the current SCM setup at Assets/Basins and highlights their deficiencies. It aims to rationalize these shortcomings by creating an SCM support group to look after the important functions such as demand forecasting, strategic procurement, VRM, and so on.

9.7.1   Creation of SCM Support Services

The structure of SCM Group in the organization has evolved over the years based on improvisation, influence of government instructions and procedures, and functioning of similar companies in the region. The SCM setup at Assets/Basins of the company is not strictly uniform—variations are observed across Assets/Basins. The variations are due to local needs, paucity or surplus resources, volume of work, size of the Asset/Basin, and so on. The current structure is not entirely based on "rationalization" or "best practices"; rather it evolved on various factors as mentioned earlier. The current SCM setup at Assets/ Basins can be broadly described as follows.

It is a flat structure with the head of SCM looking after all subgroups (departments), namely, "procurement" including purchase and contract, and "materials management and warehouse" including dispatch, receipt, stockholders, stock verification, disposal sections, and so on, and is shown in Figure 9.19.

Restructuring or reorganizing of SCM Group is beyond the ambit of this exercise, which is a subject by itself and depends on a large number of issues involving the entire organization. However, some functions that have potential of improving procurement cycle time and performance of SCM Group are found unorganized and not emphasized. The need for organizing these functions has been impressed upon.

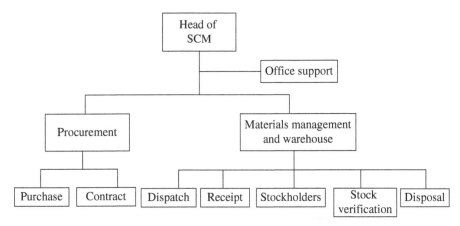

FIGURE 9.19 Existing SCM setup at Assets/Basins.

These jobs or functions are currently being done by the existing subgroups with overlapping tasks and are not clearly defined or sufficiently focused at present. There is considerable scope for improvement in functioning in terms of better coordination, clearly defined tasks, and faster response.

9.7.1.1 Suggestions: For efficient functioning of SCM support group, it would be beneficial if a subgroup is created at the headquarters (HQs) to cater to the need of all Assets/Basins to look after the following tasks:

(i) Demand forecasting
(ii) Strategic procurement
(iii) VRM

Each Asset/Basin may have a team or identified persons for these jobs depending on its size, workload, volume of transaction, suppliers concentration, and so on, who will liaise with the HQs and facilitate these functions in the Asset/Basin. The details of activities and tasks of this subgroup/Team are described in the following paragraphs.

9.7.2 Demand Forecasting

As explained in Section 9.5.2, the demand forecasting and material planning are done manually at present. The annual material requirement of each unit is collected through paper requisition and consolidated and processed for ordering each year. It is a time-consuming and tedious process resulting in long lead time. It is an inefficient process for forecasting the demand of a large number of items. This process can be improved upon as **suggested** in the following text:

(a) A team may be created at the HQs for demand planning and focal persons may be identified at the Asset/Basin depending on the workload of the SCM support group and size of the Asset/Basin. The historical material consumption data coupled with future work plan would help in generating (system-based) demand forecast of a large number of items in an efficient way in much quicker time. The generated demand may be validated by users' group, if required.

To start with, demand forecasting of "A-items" of ABC analysis or "V-items" of VED analysis may be undertaken, which would have significant impact and visible improvement within a short period.

(b) Demand forecasting and planning would also help in inventory control such as determining replenishment level, safety stock, EOQ, and so on.

9.7.3   Strategic Procurement

The company has a central procurement cell at the HQs responsible for procurement of high-value and critical materials such as drilling pipes, casing pipes, tubular, blowout preventer (BOP), drill bits, and so on, catering to the need of all Assets and Basins across the country. But the numbers of such centrally procured items are limited and pertain to drilling and well services. The following **suggestions** would improve strategic procurement:

(a) The central sourcing cell/team at the HQs may be strengthened, and more number of high-value and critical items may be procured through this cell, in order to avail the advantage of economies of scale—bulk ordering, price discount, and so on.

(b) Establish long-term contract for more number of centrally procured items with staggered delivery to different locations/regions. This would reduce the number of transactions, processing workload, manpower requirement, and above all the procurement lead time and inventory level.

(c) Demand forecasting database may be integrated with the strategic procurement system or made available to the strategic procurement team for quickening the process of procurement.

9.7.4   Vendor Relationship Management

The concept of VRM is nonexistent or largely neglected in the organization. There is no formal or planned activity on this, although vendors' meeting is organized sometime on ad hoc basis in some Assets or region (conglomerate of Assets/Basins located in the close-by geographical area).

The introduction of following major activities related to VRM would help in reducing procurement cycle time to a great extent: (i) prequalification work and selection of vendors, (ii) vendor performance evaluation, and (iii) strategic partnership.

9.7.4.1   Prequalification and Selection of Vendors: The prequalification work is presently done for a limited number of items. Most of the procurement is made through open tender following the government-specified rules and regulations. There is no formal or established team or identified persons for these works, which are done occasionally as and when required. The following **suggestions** would improve the prequalification work:

(a) In order to have compatibility with the "best practices," the prequalification work may be entrusted to a dedicated team at the HQs who will be responsible for market research and intelligence for identifications of new sources of supply and technologies.

(b) Prequalification work is largely associated with the market intelligence and external information, which may be difficult for in-house professionals to acquire. Therefore, the services of external agencies such as credit rating agencies may be availed.

(c) Prequalification database may be integrated with the procurement system for making it a robust system. All these would help in reducing the procurement cycle time of the company.

(d) Further prequalification and selection-related tasks and suggestions have been enumerated in Section 9.5.3.1.2 (h–k).

9.7.4.2 Vendor's Performance Evaluation: Furthermore, the practice of assessing suppliers' performance or rating is not followed or done at present. However, blacklisting of vendors is done for breach of contract or failure of honoring the commitment. Periodic evaluation of vendors' performance and rating them is an important step and is a best practice measure followed by IOCs. This may be introduced in the organization under reference.

9.7.4.3 Strategic Partnership: Strategic partnership with reputed and reliable vendors is beneficial to both parties. It helps in reducing procurement cycle time as well as procurement cost. Therefore, fostering strategic partnership with trusted vendors may be undertaken and encouraged.

The following suggestions would strengthen VRM:

- Establish a VRM team at HQs and Assets to look after the following jobs:

 (i) To improve communication and building confidence with suppliers

　(a) For timely delivery of materials

　(b) To reduce delivery time

　(c) To comply with best practices

　(d) To promote transparency

 (ii) Periodic measurement of vendors' performance

 (iii) Fostering strategic partnership

9.8 BENEFITS

The procurement cycle time can *be reduced by (at least) 6 months*:

Procurement cycle time = material planning time + order processing time + delivery time
$$(14\text{–}17\ months)\qquad\qquad (4\ months)\qquad\qquad (5\tfrac{1}{4}\text{–}8\tfrac{3}{4}\ months)\qquad\qquad (4\ months)$$

The average order processing time varies from $5\tfrac{1}{4}$ to $8\tfrac{3}{4}$ months against the company norm time of $2\tfrac{1}{2}$–$3\tfrac{3}{4}$ months. In other words, it exceeds the norm time by 75–166%.

The average order processing time for many items is observed at $5\tfrac{1}{4}$ months against the company norm time of $2\tfrac{1}{2}$–$3\tfrac{3}{4}$ months. Therefore, it may be presumed that following suggestions mentioned in Sections 9.5.2.2, 9.5.3.1.2, 9.5.4.2, 9.5.5.2, 9.5.6.2, 9.7.2, 9.7.3, and 9.7.4, the average order processing time may be reduced from $5\tfrac{1}{4}$–$8\tfrac{3}{4}$ months to 4 months—a reduction of 4 months.

Further, it is possible to reduce the current material planning process time (3–4 months) at least by half (i.e., 2 months) by following suggestions in Section 9.5.2.2 such as (i) dispensing manual (material) planning process for regularly consumed items, (ii) integrating material planning process with annual budgeting process, and (iii) eliminating case-by-case approval, introducing standard format for uniform and speedy approval, introducing rate contract, and so on.

Thus, at conservative estimate, procurement cycle time can be reduced by 6 months. *In other words, reduction of inventory by 6 stock-months is a modest target, which would bring the inventory level to that of 3 years back* (Year $N-3$; refer to Section 9.6.1.1b, Fig. 9.14). This is not an aggressive target compared to the benchmark of similar NOCs, regional E&P companies, and above all the company's own norm (refer to Figs. 9.2, 9.4, and 9.5). Therefore, a reduction of inventory by 6 stock-months would result in a **saving of USD 97 million** per year at offshore Assets alone and **USD 244 million** per year across the organization (based on apportioning of inventory value in terms of stock-month), keeping in view the long-term rate contract (2–3 years) with staggered delivery and exercising right of the company to modify intervals of staggered delivery on need basis. Apart from the aforementioned, the study prescribes improvement in process and functioning of some key supply chain functions.

9.9   THE SCOPE FOR FURTHER WORK

This work may be further enriched by studying and analyzing "spend analysis" more intensively. This could not be done as sufficient data were not available for this purpose. The ERP system in the organization was under stage-wise implementation, and there were issues with data migration and integration. Spend analysis for both purchase and contract would help in the following:

(a) Identifying top suppliers, which, in turn, would facilitate in suppliers' segmentation. This (suppliers' segmentation) would aid in focusing on limited number of capable suppliers with proven track record leading to better supplier relationship management. Further work in this area is required to be done.

(b) Identifying top category materials, which would help in formulating category strategy. Although long-term agreement (blanket PO), consolidation of multiple POs, etc. have been recommended, further work may be carried out for improvement in category management taking into account market strategy, market intelligence, rationalization of number of transactions, cost ownership, and so on.

(c) Spend analysis would also help in (i) developing and strengthening strategic sourcing and prequalification work and (ii) determining transaction volume (PO) and rationalizing it by combining multiple POs, encouraging long-term agreements, and so on.

(d) It has been suggested to create or strengthen some important centrally led supply chain functions, namely, strategic procurement, demand planning, and VRM. Further work in these areas may be done by detailing plan for improvement.

9.10 CONCLUSIONS

The importance of SCM for smooth functioning of E&P operations needs no elaboration. It is a well-known fact that delay or disruptions in supply of high-technology tools, high-cost spares and materials, and specialized services jeopardize E&P activities costing dearly to the organization. The supply chain functions of a state-owned E&P company that is saddled with complicated rules and regulations, lengthy procedures, and excessive documentations were studied with a view to improve their functioning. The maturity of key supply chain functions was assessed using quantitative and qualitative tools and techniques and benchmarked with comparable industry standards to find out gap and potential areas for improvement.

It was felt necessary to improve procurement process and reduce procurement cycle time by streamlining tendering process, order execution, material categorization, and so on, as these were found much higher than comparable industry standards. It has been suggested to improve the work of materials management group, namely, optimize inventory level/cost and improve warehouse functions. Furthermore, some SCM support services, which have potential of improving the SCM system, are found unorganized or nonexistent in the organization. The need for strengthening these support functions, namely, demand forecasting, strategic procurement, and VRM, have been impressed upon. Finally, corrective measures to overcome the current shortcomings and improve functioning of the SCM system as suggested in this study are summarized as follows:

1. In order to improve material planning process and reduce cycle time, the following suggestions are made:
   (i) The current manual material planning process and demand forecasting based on paper requisition may be replaced with system-generated consumption standard validated by the user group for regularly consumed materials.
   (ii) Integrate material planning process with annual budget, and dispense with case-by-case approval.
   (iii) For speedy budget and approval process, introduce standard format/checklist containing specific information, and use email for obtaining BQ, technical clarifications, and so on.
2. In order to minimize procurement cycle time, the following measures may be taken:
   (i) Encourage rate contract for more number of regularly consumed items for a longer duration with staggered delivery. This would eliminate repeated tendering.
   (ii) Enforce strong prequalification criteria for empaneling reputed manufacturers and suppliers. Identify reliable suppliers for various items and invite them to participate in open tender for empanelment. Introduce suppliers' performance and rating, and review it annually.
   (iii) Float limited tender among the empaneled vendors for procurement of regularly consumed items ensuring quality and delivery schedule and availing discount for bulk and longer-duration purchase.
3. Consider waiving of bid bond, tender fee, performance guarantee, and so on, for empaneled vendors. Introduce supplier's performance measurement and rating.

4. Encourage e-procurement; analyze lack of competition and act on it accordingly.

5. Simplify voluminous bid documents, and review terms and conditions such as bid bond, tender fees, LD clauses, payment terms, and so on, especially for low-value tenders.

6. In order to ensure best technical and commercial offer, enforce stringent technical qualification criteria for participation in open tender, so that only reputed and capable companies participate and qualify. This would check undercutting by unknown or unreliable companies.

7. Common items in the region (i.e., Assets/Basins situated in close proximity and/ or nearby geographical area) may be procured centrally.

8. Encourage rate contract and O&M contract with OEMs or authorized supply houses for supply of spares, technical services, overhauling of equipment/machineries, and so on, for longer duration with agreed terms and conditions including provision for discount, escalation of prices, and so on.

9. Authorized workshop (mechanical, electrical, instrumentation, and automobile) may be empaneled in each work center/Asset/Basin, as per need for work-up to a specified value with agreed rates, terms, and conditions.

10. In order to minimize procurement cycle time, streamline processes, and reduce administrative work, it is suggested to

    (i) Separate procurement of core and noncore businesses based on their functionality and criticality, depending on the workload, volume of transaction, and size of Assets/Basins

    (ii) Encourage long-term contract/blanket PO for non-items with low value and short lead time

    (iii) Consolidate POs for the same vendors, develop schedule for PO release to top vendors, and so on

11. User group ought to be responsible for excess buildup of inventory, whereas materials group is responsible for the inventory in transit. Efforts may be made for liquidation of MIT throughout the year.

12. For maintaining optimum inventory, reduced lead time, and faster response, introduce system-generated replenishment level/order and safety stock with validation from user group in accordance with future work plan. Automate system integration between materials management and "maintenance and operations" planning.

13. The list of surplus materials of each Asset/Basin may be circulated regularly (biannually/annually) across the organization for possible transfer. The rest may be disposed off following company rules.

14. Create a central subgroup (namely, SCM support services) at HQs with suitable representation at Assets/Basins to look after the following tasks, namely, (i) demand forecasting, (ii) strategic procurement, and (iii) VRM, which are currently unorganized and done on ad hoc basis with overlapped responsibilities. The suggestions emphasizing the need for improving these activities would help in reducing procurement cycle time and material delivery time, improving inventory management system, building confidence with vendors, and complying with best practices.

    (Refer to Sections 9.5.2.2, 9.5.3.1.2, 9.5.4.2, 9.5.5.2, 9.5.6.2, 9.6.1.2, 9.6.1.3.2, 9.6.2.1.2, 9.6.2.2.2, 9.7.1.1, 9.7.2, 9.7.3, and 9.7.4 for detailed suggestions.)

9.10.1 Benefits

The previously mentioned suggestions would help in improving key supply chain functions and would reduce the procurement cycle time at least by 6 months and thereby inventory by 6 stock-months, which is a modest target compared to similar E&P company benchmark or even its own norm. This would entail a saving of USD 97 million per year at offshore Assets alone and USD 244 million per year across the organization at a conservative estimate.

Chapter 10 deals with manpower optimization emphasizing on multiskill and multidisciplinary approach and aligning with the best practice benchmark. A case study illustrating various methods and aspects of strategic workforce planning has been discussed at length. It includes manpower demand forecasting, supply predictions, balancing and rationalization of manpower, and related issues.

REVIEW EXERCISES

9.1 Describe the evolution of SCM.

9.2 As a supply chain specialist, how would you rationalize procurement and material planning process and reduce procurement cycle time in your organization? Give some examples on areas of concern and suggestions for improvement.

9.3 What are the shortcomings of tendering process in your organization? As a supply chain specialist, what are your suggestions for improvement?

9.4 What are your suggestions to improve the procurement process and lead time of spares in your organization?

9.5 What are the advantages of categorization of materials? As a materials manager, what are your suggestions for introducing it in your organization?

9.6 As a materials manager, what are the steps you would take to optimize inventory and improve warehouse functions in your organization?

9.7 What are your suggestions for effective disposal of surplus materials in your Asset/business/organization?

9.8 What is the status of the following functions in your organization? (a) Vendor relationship management, (b) demand forecasting, and (c) strategic procurement. What are your suggestions to improve SCM support group in your organization including the aforementioned functions?

REFERENCES AND USEFUL LINKS

[1] Simchi-Levi, D., Kaminisky, P., and Simchi-Levi, E., Designing and Managing the Supply Chain: Concepts, Strategies, and Case Studies, Irwin/McGraw-Hill, Boston, MA, 2000.

[2] Materials Requirement Planning: Wikipedia, available at: http://en.wikipedia.org/wiki/Material_requirements_planning (accessed on March 2, 2016).

FURTHER READING

Bozarth, C. and Handfield, R., Introduction to Operations and Supply Chain Management, Prentice-Hall, Upper Saddle River, NJ, 2006.

Chopra, S. and Meindel, P., Supply Chain Management: Strategy Planning and Operations, Pearson Prentice-Hall, Upper Saddle River, New Jersey, 2002.

Garg, V.K. and Venkitakrishnan, N.K., Enterprise Resource Planning: Concepts and Practice, 2nd edition, Prentice-Hall of India, New Delhi, 2003.

Handfield, R.B. and Nochols, E.L., Jr., Introduction to Supply Chain Management, Prentice-Hall, Upper Saddle River, NJ, 1999.

Hines, T., Supply Chain Strategies: Customer Driven and Customer Focused, Elsevier Publishing, Oxford, 2004.

Hopp, W.J. and Spearman, M.L., Factory Physics: Foundations of Manufacturing Management, Irwin/McGraw-Hill, Boston, MA, 1996.

Malakooti, B., Operations and Production Systems with Multiple Objectives, John Wiley & Sons, Inc., Hoboken, NJ, 2013.

Monk, E. and Wagner, B., Concepts in Enterprise Resource Planning, 3rd edition, Course Technology Cengage Learning, Boston, MA, 2009.

Ptak, C. and Smith, C., Orlicky's Material Requirements Planning, 3rd edition, McGraw-Hill, New York, 2011.

Shapiro, J.F., Modelling the Supply Chain, Duxbury Thomson Learning, Pacific Grove, 2001.

Sheilds, M.G., E-Business and ERP: Rapid Implementation and Project Planning, John Wiley & Sons, Inc., New York, 2001.

Sumner, M., Enterprise Resource Planning. Prentice Hall, Upper Saddle River, NJ, 2005.

Sweeney, E. (ed.), Perspectives on Supply Chain Management and Logistics: Creating Competitive Organizations in the 21st Century, Black Hall Publishers, Dublin, 2007.

Tayur, S., Ganshan, R., and Magazine, M. (eds.), Quantitative Models for Supply Chain Management, Kluwer Academic, Boston, 1999.

Viswanadham, N. and Narahari, Y., Performance Modeling of Automated Manufacturing System, Prentice-Hall of India, Upper Saddle River, NJ, 1998.

Useful Links

Enterprise Resource Planning: Wikipedia, available at: http://en.wikipedia.org/wiki/Enterprise_resource_planning (accessed on March 2, 2016).

Manufacturing Resource Planning: Wikipedia, available at: http://en.wikipedia.org/wiki/Manufacturing_resource_planning (accessed on March 2, 2016).

Supply Chain Management: Wikipedia, available at: http://en.wikipedia.org/wiki/Supply_chain_management (accessed on March 2, 2016).

10

MANPOWER OPTIMIZATION AND STRATEGIC WORKFORCE PLANNING

10.1 INTRODUCTION

In earlier chapters, we have dealt with optimization of cost, strategy, and performance of various operational and functional areas. In this study, we would deal with manpower optimization and workforce planning, which is an important issue in large enterprises with tremendous influence on organizational productivity. Manpower utilization, development, and cost have significant bearing on large companies, as excess manpower breeds inefficiency and deficient manpower leads to the loss of opportunity and productivity. Therefore, it is important that enterprises have optimum workforce to meet their strategic goals. There should not be a wide gap between requirement and availability of manpower that might affect the work program, strategic goals, and objectives of the organization. A lean organization essentially has a lean workforce!

The primary purpose of manpower planning is to match the requirement (number) of people with specific skill sets at a specified time for a stated level of activities at an optimum cost. The oft-repeated text describing manpower planning—"To provide right number of people with right qualification at the right time"—is true for ideal condition (utopian state). But in reality, there are many constraints and perfect match is rather difficult. It is important that manpower plan is aligned with the strategic goals and objectives of the organization and shall not be guided by short-term need. Many organizations adopt ad hoc plans to tide over their immediate manpower need without considering their gainful utilization over the long run. Such practices are detrimental in the long run and should be discouraged, as it is easier to regulate induction than terminate the job later (which has far-reaching consequences and is elaborated in Section 10.5).

There are many facets of manpower planning. The current study deals with some key functions of workforce planning including demand forecasting, supply predictions, and

Optimization and Business Improvement Studies in Upstream Oil and Gas Industry, First Edition.
Sanjib Chowdhury.

strategies for balancing manpower demand–supply gap. It discusses various aspects and methods of demand forecasting, identifies potential flaws in manpower planning process, and suggests safeguards against these. The study impresses on multiskilling and multidisciplinary approach to improve manpower utilization and productivity and enrich the quality of the human resources (HR). It identifies low-utilization categories and disciplines by conducting work sampling and rationalizes existing disciplines, subdisciplines, and categories in the organization under reference, whose numbers are found far in excess of the best practice benchmark. It is important to note the difference between discipline, subdiscipline, and category referred to in this study, which is illustrated in the following table:

Discipline	Subdisciplines/Trades	Category
Mechanical engineering	Mechanical	Sr.-level mechanical engineer
	Machining	Mid-level mechanical engineer
	Fitting	Jr.-level mechanical engineer
	Diesel	Sr.-level/Jr.-level technicians
	Welding	(Diesel/fitting/welding/machining/
	Draftsman (mechanical)	painting/foundry…)
	Blacksmith	
	Foundry	
	Painting	

Finally, a real-life case study exhibiting optimization of strategic workforce in an E&P Asset has been deliberated at length covering important facets of manpower planning and other key issues.

10.2 OBJECTIVES

Manpower planning covers a wide range of issues that have huge influence on achieving strategic goals and objectives of an enterprise. This study tries to address some of the key issues that have great bearing on an organization's growth plan. The objectives of the current exercise are the following:

(a) To optimize strategic workforce of an E&P Asset (this has been dealt with as a case study in Section 10.8)

(b) To identify low-occupation disciplines/subdisciplines and introduce multiskilling and multidisciplinary approach for manpower rationalization

(c) To rationalize numerous disciplines/subdisciplines and trades prevailing in the organization to a manageable level at par with the industry best practice

(d) To develop and modify the organization's manning norms and staffing pattern of various operations based on reality, advancement of technology, local factors, and *alignment with the best practice benchmark*

(e) To identify excess and deficit categories and rationalize current workforce through redeployment, skill development and training, hiring, voluntary retirement, and so on.

10.2.1 Approach

The approach adopted in this exercise is described as follows:

(a) Manpower planning process of the company was studied extensively, and the shortcomings were identified.

(b) Work sampling was carried out to identify the low-utilization categories. This was supplemented with operational knowledge and long work experience in the industry. Jobs that are of similar type requiring similar skill sets or not of continuous nature are the first choice for multiskilling and multidisciplinary approach.

(c) The reasons for proliferations of disciplines and subdisciplines in the organization were found out including the effect of local factors, advancement in technology, organizational policy, and so on, and were rationalized aligning with the best practice benchmark.

(d) The methodology followed for manpower optimization has been detailed in Section 10.8.3 including processes for evaluating manpower requirement, availability, and other key issues.

10.3 MANPOWER PLANNING AND OPTIMIZATION PROCESS

Manpower planning in an E&P company is a bit more complex than conventional industries due to wide range of activities, variability in operation and technology, widely differing local or environmental factors, and so on. Manpower optimization is a multi-phased process whose main components are (i) demand forecasting, (ii) supply (availability) forecasting, and (iii) strategy for balancing them.

Demand forecasting is the heart of manpower optimization process that depends on several heterogeneous factors, such as market forces and trend, demand for the product/service, economics and financial resources, technology, organizational growth, corporate strategy, management priority, and so on. Since these factors differ widely from company to company, customized model is suggested. The techniques used in demand forecasting are (i) managerial judgment/estimate, (ii) Delphi technique, (iii) work study, (iv) job analysis, (v) productivity measurement method, (vi) best practice benchmark, and (vii) statistical methods like trend analysis, regression analysis, and so on. Some of these are elaborated in detail in Sections 10.4.1, 10.4.2, and 10.4.3.

Supply forecast process starts with internal analysis of manpower inventory (i.e., checking in-house availability) and uses the following information and methods—(i) skill inventory, (ii) management inventory, (iii) succession planning, (iv) staffing tables, (v) Markov analysis, (vi) simulations, and so on. If the requirement cannot be met internally, then external sources of supply are explored, which again depend on several factors like (i) government policy, (ii) demographic changes, (iii) availability of skilled and educated workforce in the (local) market, (iv) labor migration/mobility, and so on.

Finally, the assessed manpower requirement and in-house availability are compared and strategies evolved to bridge the difference. Accordingly, recruitment plan, redeployment plan, retention plan, redundancy plan, training, and other plans are developed and implemented.

The shortages are generally met through (i) hiring full-time, part-time, or contract employees; (ii) internal transfers; (iii) promotions; (iv) training; (v) paying overtime; and so on. The surpluses are managed with (i) training and redeployment in the areas of need, (ii) internal transfers, (iii) voluntary retirement, (iv) retrenchment in extreme cases, and so on. All these are elaborated with real-life case study in Sections 10.8.5, 10.8.6, and 10.8.7.

10.4 METHODS OF MANPOWER DEMAND FORECASTING

Manpower forecasting methods that are generally followed in E&P companies are *(i) assessment based on the best practice benchmark and (ii) assessment based on manning norms and work practices followed by the enterprise. Apart from these, there are (iii) mathematical optimization models and technique that are rarely used in E&P sector.*

These methods are described in the context of the current manpower optimization study and presented in the following paragraphs.

10.4.1 Demand Forecasting Using "Best Practice Benchmark"

Determining manpower requirement based on the "best practice benchmark" is a progressive method and is widely followed in E&P industry wherein the organization under study is compared with one or more similar companies that follow the best practice. Accordingly, manpower deployment pattern for each unit, activity, and operation is drawn and consolidated with provisions for legal or statutory requirement. The important issue is the selection of an appropriate benchmark company (or companies) that is (or are) considered similar to the company being assessed.

The critics point out that an E&P company in a developing or underdeveloped country cannot be compared with that of a developed country because of wide differences in the maturity level of local market, infrastructure, facilities, and services. Even within a vast country, the operating conditions in different regions may vary considerably in terms of the aforesaid factors, besides work culture and practices. *Moreover, the labor laws, employment rules, and regulations of the state also influence employability plan of an E&P company. An Asset or work center in a developed country or region may outsource many activities and services, which are not available in remote, inaccessible, and undeveloped region. Therefore, an Asset in these undeveloped areas has to develop and manage all these activities and services of its own, which would require huge investment and additional manpower.* These may make the comparison inappropriate.

For example, most of the IOCs and NOCs in the developed countries/regions outsource many services, such as drilling, cementing, logging, seismic survey, shot hole drilling, well services, repair and maintenance of equipment (mechanical, electrical, instrumentation, and auto), logistics, housekeeping, civil works, civil maintenance, and many others, which may not be available or partially available in underdeveloped countries or regions. As a result, many of these services are being maintained and managed in-house by many enterprises in the underdeveloped or developing countries. Thus, there will be dissimilarity in requirement of manpower, and comparison on this count will be inappropriate.

There may not be an apple-to-apple comparison, but these cannot be the perpetual excuse for not adopting the best practice benchmark. The conditions prevailing 50 years back in the country are not the same now; it has progressed a lot and the world has become more globalized. Therefore, the best practice benchmark shall be the guiding factor for setting manpower norms in E&P companies. This is not an insurmountable problem to tackle, although there may be variances depending on local factors and legal or statutory requirements in different countries or regions.

10.4.1.1   Suggestion: Based on the observations, discussion, and analysis in Section 10.4.1, it is suggested that:

> An E&P company strives for the best practice benchmark, notwithstanding the level of maturity of local market, infrastructure, facilities, and services available. The best practice benchmark with some allowances for local factors and legal or statutory requirements may be adopted as manning norms of the company operating in underdeveloped and developing countries or regions. The local factors and legal or statutory requirements may vary from country to country or region to region depending on the aforementioned factors.

10.4.2   Demand Forecasting Using Manning Norms and Work Practices

Manpower requirement is evaluated based on organization's manning norms and work practices. Many E&P companies have developed their own manpower norms and work practices for various technical operations and nontechnical activities. These were developed over the years based on the work culture, in-house work study, discussion, and assessment. Most of the technical operations like production installations including GGS, CTF, and EPS; drilling and related operations like fishing, DST, rig move, and cementing; well services containing workover operations, production testing, CTU, and well stimulation; logging; mud chemistry; and so on have established manpower norms, which is also influenced by local factors to some extent. But the same for office jobs are usually not so well defined and are often *influenced by judgment rather than the actual assessment of workload*. This is true especially for nontechnical disciplines like HR, finance, supply chain management, general management, administration, and so on.

10.4.2.1   Dangers   of   Judgment-Based   Assessment   Judgment-based   manpower assessment is generally done for office/desk jobs, which are difficult to quantify. It is not a good practice and has serious flaws, as perception and judgment vary from person to person and expert to expert and even among groups of experts. It suffers from prejudice, as the so-called experts who are associated with the respective department for decades have silo-outlook. Judgment of the boss rules the roost for such type of assessment. The higher the position or authority of the boss, the greater is the chance that his judgment would prevail. Moreover, *judgment-based assessment often degenerates to some sorts of bargaining and is influenced by personal relations and other unrelated factors*. Few real-life examples are cited in the following paragraphs.

Example 10.1

Once, an influential senior executive of an E&P company was clamoring for more manpower for his Asset. He was forcefully placing his contentions before the Director–HR of the company, who was not in agreement with the demand of the senior executive of the

Asset. After a year or so, the same senior executive became the Director–HR of the company. The same demand that he had asked for earlier was drastically pruned and approved by him as Director. Perception and judgment change with the change of side or chair!

There are many instances where additional posts or manpower is allegedly approved based on personal relationship between the approving authority and the concerned group or department heads, even though there are no changes in the current as well as future work plan, activities, and workload. This may sound strange but not uncommon! It is observed in many organizations and cannot be brushed aside as isolated cases.

Example 10.2

On another occasion, a region, that is, conglomerate of five large Assets in nearby geographical areas, was clamoring for additional manpower from the corporate headquarters. It had put up its assessed requirement—an additional 2000 employees in various disciplines to cope with increased activities in the near future. Manpower assessment team from the headquarters headed by a senior executive visited the region for review and finalization of manpower demand of different Assets in that region. Several rounds of high-level meeting were held between the headquarters team and the senior officials of the region for few days. There were vociferous demands for additional manpower from different Assets; arguments and counter-arguments followed. At the end of the marathon meetings, an important senior executive of the region confided to his team, "Out of 2000 people we have asked for, if we get 40 odd people that would suffice the requirement of all Assets in the region." So, one can imagine the danger of judgment-based assessment, which looks like bargaining to some extent. Such type of assessment should be replaced with the assessment based on the actual workload.

Example 10.3

There are many potential flaws in assessing manpower demand, and one should take adequate care while carrying out such an exercise. For example, some companies have rolling 3- or 5-year manpower plan, which is updated every year, and annual manpower budget is linked with it. All groups/units are requested to submit their manpower demand every year for preparation of annual manpower budget. *The process apparently looks elegant and seems like a good practice. But in reality all groups/units provide inflated requirement with superfluous justifications devoid of in-depth assessment of workload. These are routinely processed and the practice is not free from prejudice.*

Such type of manpower assessment based mostly on judgment is misleading. This should be replaced with proper assessment by experts/consultants based on (i) best practice benchmark or (ii) sound organization norms or (iii) actual workload and taking into account the cost–benefit analysis.

There is a tendency in many organizations to have some extra manpower for safeguarding from uncertainties. There may be genuine shortages of manpower in few disciplines or categories, which are projected repeatedly, giving the impression that there are massive shortages in the entire group/unit/Asset. These shortages become the rallying point for additional manpower without considering that there are excesses in other categories within the same department or group. The possibility of utilizing surplus manpower to mitigate shortages should be explored, and the existing strength should be rationalized to the extent possible. *These excesses are the result of improper assessment of demand fore-casting, short-sighted induction policy, skewed promotion policy, and so on.* The case study presented later in this chapter demonstrates this conundrum, which may be noted in other organizations.

10.4.2.2   Suggestions: Based on the observations and deliberations in Section 10.4.2, the following suggestion is made:

> The manning norms of an enterprise shall not be based on judgment but on the actual assessment of workload. It should not vary widely with that of the best practice bench-mark, although some allowances for local factors and legal or statutory requirements are permissible.

10.4.3   Mathematical Optimization Models and Techniques

Mathematical optimization models and techniques look elegant, but its applicability is limited to few industries where uncertainty and variability are much less and assumptions are relatively simple. But E&P companies operate under high degree of uncertainty and variability, which make mathematical models somewhat complex, and results are usually less reliable. Generally, such models have more academic value than use in real life.

Mathematical models that are normally used for manpower planning are linear program-ming, integer programming, scheduling and assignment problems, and so on. But their use in real life is limited to specific industry as mentioned before. The complexity involved in mathematical modeling and the usefulness of results (which are generally known to the experienced professionals of the respective industry/enterprise and system) make it less popular.

On the other hand, best practice benchmark is widely used and is the popular choice of management consultants, which allows unit-based assessment of activities and provides accurate estimation of manpower demand.

10.5   OTHER KEY ISSUES IN MANPOWER PLANNING

The outcome of manpower demand forecasting and supply analysis is the basis of recruitment, which requires careful scrutiny and sound planning. It should be guided by the long-term business plan of the company, as recruitment of one person means carrying him for the next three decades or so. Downsizing is a sensitive issue, especially in the NOCs/state-owned companies, where retrenchment is difficult due to government rules and regulations. Besides, ruthless retrenchment invites serious industrial relations problem, especially during the downturn of the economy. Such action is not a good option in this era of cyberage. It does more harm than good as negative news is dissem-inated faster than ever before, thus severely tarnishing the reputation of the company. Therefore, *it is important to recruit people in a phased manner based on judicious assessment of manpower demand keeping in mind the long-term strategic goal of the organization.*

Effect of Organization Structure: Manpower requirement of an organization/unit/group also depends on its structure. Therefore, it is important that the respective entity has appropriate structure in place and follows the right business model. *The structure shall essentially be lean and conducive to efficient work and processes. The compartmentaliza-tion of work should be discouraged, and multidisciplinary approach may be followed for optimum utilization of manpower. The expansion of organization structure and increase in manpower should justify the cost–benefit analysis. It must lead to improved efficiency, focus on core activities, optimum span of control, and so on.* However, each organization and operating environment is unique, which is required to be judged on its own context taking into account local factors and long-term goal of the organization.

Apart from matching the demand–supply gap, it is important to focus on the **quality of HR**. In this era of competition, the quality of manpower is a distinguishing factor that gives an edge to an enterprise over competitors. In fact, an organization is better known and admired for its quality of HR, which also augments its brand value.

Manpower cost is a determining factor for profitability of many organizations. It includes salary, allowances, perquisites, welfare cost, and the like, which are usually met from the OPEX budget. Besides, there are indirect cost of recruitment, training and development, maintaining the workforce at a desired efficiency level. There are hidden costs like under-utilization of manpower and carrying them over the years, job turnover, and so on. All these aspects are to be kept in mind while carrying out the exercise of manpower planning.

10.6   INTRODUCTION OF MULTISKILLING AND MULTIDISCIPLINARY APPROACH

There are many categories and disciplines whose staffing norms are disproportionate to their utilization. For example, there is plenty of maintenance crew in a deep drilling rig with seemingly low utilization. As per work practices of the company, there are two sets of maintenance crew in the deep drilling rig, namely, mechanical maintenance and electrical maintenance crew in each shift. Electrical maintenance crew in each shift consists of four personnel—one electrical engineer and three technicians (electrical). Similarly, mechanical maintenance crew in each shift consists of four to six personnel—one mechanical engineer and three to five technicians (diesel/fitting)—depending on the type and capacity of drilling rig. This staffing norm is being followed since the inception of the company and has not been modified for various reasons.

As per the work practice of the company, the job of rig maintenance crew is restricted to checking lubricants, monitoring machineries, undertaking minor repairs of rig equip-ment, and so on. In case of medium and major breakdown, the field maintenance party stationed at the base is called for, as rig maintenance crew is not capable of repairing these. The utilization of rig maintenance crew is apparently low, but the trade unions/ employee associations do not agree to this; notwithstanding, the company staffing norm is much higher than that of the best practice benchmark. Expectedly, there was stiff resis-tance for changing manning norms from various quarters including the collectives (trade unions/employee associations). In order to overcome this, the activity or work sampling was carried out to find out the actual utilization of maintenance crew at the deep drilling rig, which is elaborated in the following paragraph.

10.6.1 Identification of Low-Utilization Trades/Disciplines/Categories

- The **purpose** of this exercise is to determine the occupation time of electrical and mechanical maintenance crew at the deep drilling rig.

10.6.1.1 Approach: The following approach was followed:

(a) Activity or work sampling that is based on statistical theory was adopted for the study. The exercise was carried out at an electrical rig (1500 HP capacity) at 95% confidence level with adequate sample size (600 nos.). The observations were made at random over a period of 1 week.

(b) An ad hoc allowance of 10% (personal—4%, fatigue—3%, and delay and contingency—3%) was considered to determine the occupation time. Percentage working include "worker working," "worker inspecting," and "worker engaged"; other observation elements are "worker away" and "worker idle."

10.6.1.2 Results: The outcome of the work sampling study is presented in the following:

(a) The percentage working of electrical crew with 10% allowance varies between **22.7** and **27.3%**. The actual percentage working, upper and lower limits of percentage working (±2σ), and so on, are shown in Table 10.1.

(b) Similarly, the percentage working of mechanical crew with 10% allowance varies between **33.7** and **44.3%**. The actual percentage working, upper and lower limits of percentage working (±2σ), and so on, are shown in Table 10.2.

The summarized observation sheet with elemental breakdown and statistical computation is shown in Appendix 10.A, which would give a fairly good idea to conduct work sampling in similar circumstances.

This is a real-life exercise that can be replicated in other disciplines to find out employee utilization. *There are several disciplines and categories with seemingly low occupation time or whose nature of work is not continuous or works are of similar type or different crew are attending the same equipment or activity. Multiskilling and multidisciplinary approach can be adopted in such disciplines/trades/categories.* For example,

TABLE 10.1   Utilization of Electrical Maintenance Crew in Deep Drilling Rig

Crew	% Working	Upper and Lower Limits of % Working (±2σ)	% Working with 10% Allowance
Electrical engineer	17.3	24.8 9.8	27.3
Technician #1	15	22.1 7.9	25
Technician #2	12.7	19.3 6.1	22.7
Technician #3	15.7	22.9 8.5	25.7

TABLE 10.2   Utilization of Mechanical Maintenance Crew in Deep Drilling Rig

Crew	% Working	Upper and Lower Limits of % Working (±2σ)	% Working with 10% Allowance
Mechanical engineer	26.1	34.8	36.1
		17.4	
Technician #1	24.7	33.3	34.7
		16.1	
Technician #2	23.7	32.2	33.7
		15.2	
Technician #3	34.3	43.7	44.3
		24.9	
Technician #4	Occupation time of these two crew members could not be		
Technician #5	determined due to inconsistency		

multiskilling and multidisciplinary approach may be introduced *for operating crew in production installations*. Similarly, it can be adopted *for maintenance crew in deep drilling rigs, well services, production installations, cementing services, logistics, and others.*

10.6.2   Suggestions

Based on the findings in Section 10.6.1.2, the following suggestions are made for the introduction of multitasking and multidisciplinary work and improved manpower productivity:

(a) The low occupation time of maintenance crew of the deep drilling rig suggests that there is scope for reduction of crew strength, and it may be steadily brought to the level of the best practice benchmark. For this, the maintenance crew may be trained in multidisciplinary and multitasking jobs to widen their area of operation and improve their current utilization.

(b) In order to improve their utilization, the maintenance crew may be extensively trained to repair most of the equipment and machineries at deep drilling rigs, which are currently being done by field maintenance party, thus reducing dependability on them.

(c) The role of mechanical, electrical, and instrumentation operators in the maintenance team in deep drilling rigs, well services, production installations, and others may be merged, and a "technician" category with broader role may be created.

(d) In order to identify disciplines and categories where multiskilling may be introduced, find out jobs that are not continuous in nature and/or having low occupation time or are of similar nature or trade, which can be performed with appropriate training.

(e) The operational and maintenance workers in cementing services may be merged, as they are conversant with and can do each other's job with suitable training or orientation, for example, operational crew—operator cum cementing mechanic, drilling assistant (cementing), and so on—and maintenance crew—diesel mechanic, auto mechanic, welder, and fitter. *There is little justification in maintaining two separate sets of crew for cementing units, which may be merged into one with multiskilling.*

(f) There are several categories in logistics team/department, namely, driver (light vehicle), operator (heavy vehicle), operator (crane), operator (heavy equipment), winch operator, slinger, rigger, cleaner, and so on. *These categories may be merged in one or more categories such as operator (logistics) with wider responsibility of driving, operating, and managing the vehicle and equipment at the desired level.*

(g) Multiskilling and multidisciplinary approach may be adopted for the *operating crew in the production installations.*

(h) In order to encourage the acceptability of multiskilling and multidisciplinary approach, the employees and collectives (trade unions/employee associations) may be taken into confidence, and the merged categories with broader responsibilities may be compensated suitably.

10.7 RATIONALIZATION OF DISCIPLINES, TRADES, AND CATEGORIES

10.7.1 Proliferation of Disciplines and Trades: Historical Reasons

One of the malaises of the organization under reference is that it's having more than 100 disciplines and subdisciplines, contrary to the best practice benchmark. This has led to the compartmentalization of work resulting in low occupation time and low productivity. The reasons for proliferation of such high number of disciplines are mainly due to historic and socioeconomic, environmental, and technological advancement, organizational policy, and so on, which are deliberated at length in the following paragraphs.

The company under reference was established more than 50 years back and started its operation from the grassroot level with no experience in E&P activities. The world was not so liberal at that time; there were severe restrictions in trade, technology transfer, and businesses in general. Oil- and gas-related technology and services, equipment, and materials were not readily available, which were controlled by few companies with monopolistic grip across the world. The growth in these areas was a daunting task for an underdeveloped or developing country, as technology and expertise were not forthcoming from other nations. Even if it were available, it came with a huge price, which the customers perceived as "unfair." All these led the country to adopt the policy of becoming "self-reliant" in oil and gas technology and services, promoting indigenization of materials and manufacturing of equipment. As a result, the company grew encompassing all related services, such as *drilling, cementing, mud services, well services, workover operations, logging, seismic survey, logistics, civil, repair and maintenance of equipment, and many others.*

Due to poor infrastructure and lack of industrial base, the company had to develop, manage, and maintain basic infrastructure at all work centers, such as *civil works at drill sites and installations; housing of employees at far-off work centers; mechanical, electrical, instrumentation, and auto workshops for repair and maintenance of equipment and machineries at work centers; logistics including crane, heavy and light vehicles, marine fleet, and helicopter services; housekeeping and low-end maintenance*; and so on.

The priorities of the state were different at that time; it was interested in creating more jobs, providing employment to its vast unemployed populace, and developing unemployed youths to productive resource pool. The *government rules and regulations, labor laws, employment policy, aggressive trade union movement, and political patronage* **led to the creation of more than 100 disciplines and subdisciplines in the organization**.

All these factors coupled with socioeconomic issues were responsible for absorbing thousands of unskilled and semiskilled labors who were engaged on contract and whose jobs were either seasonal or providing low-end services. Thus, one can find the post of *cleaner, helper, attendant, technical attendant (in laboratory), sanitary cleaner, helper (gardening), guesthouse attendant cum cook, record keeper, duftry, photographer, and many others.*

10.7.2 Changing World, Scope for Outsourcing, and Redundancy

Since then, the world has changed a lot; it has become more liberal and business has become more competitive than ever before. Meanwhile, the country has progressed a lot; there was phenomenal growth in industrialization, infrastructure development, and service sectors. It has become a major manufacturing giant exporting a large number of goods and offering various services to many countries across the globe. There was competition in business at home as well as abroad. Under such a situation, the rule of the game and business changed. There were pressures on the enterprise to improve operational performance and productivity, and the need for cost optimization was greatly felt. These became the necessity for survival in subsequent years.

Many goods and materials, tools and equipment, and **services**, which were earlier not available in the country, are readily available nowadays in the local market, for example, repair and maintenance of equipment including mechanical, electrical, instrumentation, and auto; logistics covering crane, heavy vehicle, and heavy equipment; housekeeping; civil works and maintenance; shot hole drilling; drilling and cementing services; logging services; and so on. *Many of these services can be outsourced nowadays for better efficiency and economic reasons. Farming out noncore activities and low-end services would allow the enterprise to focus on the core activities and strategic issues.*

With the advancement of technology and industrialization of the country, many trades and disciplines have become outdated and have no relevance in these days. As a result, under the changed scenario, thousands of employees in dozen of disciplines and subdisciplines either have become redundant or do not have requisite skills for the new technology. For example, *black-smithy, dent beating, vulcanizing and tire, Glassblowing, carpentry, masonry, plumbing, upholstery, painting, foundry, auto-electric, draftsmanship, map and drawing, photography, horticulture, shot hole drilling,* and so on, have become redundant category/discipline/subdiscipline, as these services could be obtained from the market. Retrenchment was not considered as a viable option because of sociopolitical reasons. Instead, the path of natural attrition, voluntary retirement scheme, and so on, was preferred. But these were not enough to get rid of or gainfully utilize thousands of unskilled and semiskilled workers.

10.7.3 Disciplines/Trades with Low Engagement or Occupation Time

Furthermore, the utilization of many trades/subdisciplines was found abysmally low, such as technician—*diesel, fitting, mechanical, electrical, instrumentation, refrigeration, air-conditioning,* and so on. It would be advantageous to retrain them with multiskilling and multitasking, thus rationalizing the categories/disciplines as well as the workforce.

10.7.4 Faulty Recruitment and Promotion Policy

The myopic organizational policy on recruitment and promotion was also responsible for the proliferation of disciplines and cadres in the company. This practice continued even in the comparatively matured stage of industrialization, for example, the creation of post of *data entry operator, Interpreter and translator (language), librarian, Xerox operator, and photographer, to mention a few.* These categories/cadres have restricted growth opportunities, so after few years, they were absorbed in other disciplines like HR, general management, administration, and others irrespective of their suitability.

Furthermore, the recruitment and promotion policy of the organization was found out of sync with the best practices followed by leading E&P companies. For example, the workers of all disciplines and subdisciplines in the organization can become officers on natural progression. Thus, a carpenter, mason, plumber, and data entry operator can become an engineer/officer in due course of time without enhancing their qualification or skill sets. Therefore, the entry to officer cadres may be restricted, which shall be guided by the induction-level criteria for that post and not on natural progression.

10.7.5 Rationalization of Disciplines/Trades

All these led to rationalization of mammoth workforce of the company and more importantly to rationalize the huge hordes of disciplines/subdisciplines to a manageable level. There were little efforts or will to rationalize these vast caches of categories in all these years mainly due to sociopolitical reasons and collective (trade unions/ employee associations) pressure. Therefore, there is an urgent need for rationalizing the huge number of disciplines/subdisciplines (around 100) prevailing in the organization at present and bringing it down to a manageable level of around two dozen as per the industry practice. Many in-house taskforce and committees were set up in the past to rationalize this large number of disciplines/subdisciplines/categories, but these were done in piecemeal, and the scope of these studies was limited to few operational areas.

10.7.6 Suggestion

Based on the aforementioned reasons, analyses, and findings, the following suggestions are made to rationalize large number of disciplines and categories to the optimum level:

(a) In view of the advancement of technology, greater scope of outsourcing noncore activities, and outdated and redundant subdisciplines under changed conditions, there is a need to rationalize disciplines/subdisciplines/categories and improve the efficiency of the workforce. Therefore, it is suggested to abolish the following disciplines/subdisciplines/categories:

Civil:
- Plumbing
- Mason
- Carpentry

- Draftsman
- Surveying

Workshops:
- Blacksmith
- Dent beating
- Vulcanizing and tires
- Painting
- Auto-electric
- Auto
- Upholstery
- Glassblowing
- Foundry
- Refrigeration and air-conditioning

Office Jobs:
- Record keeping
- Xerox operator
- Duftry
- Map and drawing
- Data entry operator
- Photographer
- Interpretation and translation (language)
- Steno

Others:
- Horticulture
- Mali (gardening)
- Guesthouse attendant cum cook
- Crane
- Heavy vehicle
- Heavy equipment
- Shot hole drilling

Medical Services:
- Attendant cum dresser
- Pharmacist
- Nurse
- Matron and various other categories

Marine Discipline:
- Foreign going
- Marine radio and other categories

Unskilled Workers:
- Cleaner
- Sanitary cleaner

- Helper
- Attendant

The aforementioned list is illustrative only.

(b) Based on the earlier deliberations and best practices followed in similar E&P companies, it is suggested that *the large pool of current disciplines/categories may be consolidated to a manageable level of around two dozen in line with leading E&P companies.* An illustrative list of proposed disciplines for the organization under study is shown as follows:

- Geoscience (geology, geophysics, petrophysics, etc.)
- Reservoir engineering
- Production
- Drilling/well operations
- Electrical
- Mechanical
- IT and telecommunication
- Civil and construction
- HR and general management
- Finance and accounts
- Quality health safety environment
- Legal
- Supply chain management
- Business development, planning and strategy, and others

10.8 A CASE STUDY ON MANPOWER OPTIMIZATION

10.8.1 Background

This study was carried out at a remote onshore Asset, which is far away from the mainland and whose dominant mode of connectivity is by air. The land route is tortuous and severely constricted. The entire region is underdeveloped in terms of infrastructure, industrialization, and availability of various services but has potential for growth. The region is rich in natural resources with abundance of natural gas and extensive forestation. Gas reserves were discovered three and half decades back; since then, further exploration and limited production operations are ongoing. The full potential of gas production of the Asset has not been realized, as there are a limited number of customers and industries in the region. The investment made by the Asset/E&P company is locked up in the subsurface for a long time, and there is a need to monetize the asset. The development and industrialization of the region are intrinsically dependent on the utilization of locked-up underground gas reserves. It was decided to build a gas-based power plant, which would help tide over the growing demand for power and propel the economic growth of the region.

But this would necessitate the augmentation of gas production and handling capacity from the present approximately 75–300 MMSCFD, that is, fourfold increase of the current gas production and handling capacity. This would also require increase in all-round activities in various disciplines, which are enumerated in Section 10.8.4.

It was observed that some departments/groups were supposedly facing shortages of manpower in certain categories and disciplines. The concerned department/group heads were clamoring for additional manpower in these categories. Soon, all department/group heads felt that they were understaffed for the current level of activities, and this would aggravate with increasing activities in the near future. They voiced their concern for additional manpower, which led to the assessment of manpower demand and manpower optimization study of this remote Asset.

10.8.2   Objectives

In order to achieve the strategic goals and objectives of the company, the activities of the Asset is expected to rise manifold, especially in the areas of drilling, production, handling and supply of gas, support services, and other related works. This would necessitate appropriate manpower in terms of competency, skill sets, and quantity (number). The objectives of this study are as follows:

(a) Optimize strategic workforce in an E&P Asset taking into account (i) current and (ii) future activity level
(b) Identify surplus and deficit categories and rationalize deployment of existing workforce
(c) Formulate strategy to overcome shortages and minimize surplus manpower through various means including gainful utilization or otherwise

10.8.3   Methodology

The methodology followed in this study is described as follows:

(a) The structure of the Asset, chain of command, existing job contract, service contract, local factors, and so on, was studied, as these have considerable influence on the manpower requirement of the organization.
(b) Manpower demand forecasting was made based on both (i) the best practice benchmark and (ii) the company norms and work practices and taking into account the workload, future work plan, local factors, and so on.
(c) Manpower requirement using both these methods were compared and suggestions offered to improve organization's manning norms and work practices so as to align with the best practice benchmark.
(d) Extensive discussions were held with all stakeholders and department/group heads to know their views on various issues including contentious ones.
(e) The current manpower inventory data of the Asset were captured, and both quantitative and qualitative analyses were carried out to know the skill sets, qualifications, competency, age profile, personal traits, constraints, and so on.

(f) The manpower demand was assessed and compared with the existing manpower in the Asset. Accordingly, surplus and deficit categories were identified. Rationalization of manpower to the extent possible was done through mobilizing the excess categories to mitigate the deficiency of the similar skill sets and category.

10.8.4   Activities and Facilities

The current as well as future activities, workload, and facilities considered for workforce planning of the Asset are shown in Table 10.3.

The activities and workload of the previously mentioned groups/departments/teams are furnished in the respective assessment sheet in Appendix 10.E.

TABLE 10.3   Current and Future Activities and Facilities

Activities and Facilities	Year N (Current)	Year N+3
Drilling rig	Owned—3 nos. (28 days on/off pattern) Drilling of 6–7 wells per year (average depth 12,000 ft)	Owned—3 nos. Hired—2 nos. drilling of 10–12 wells per year for the next 10 years
Production installations	GCS—4 nos. Gas production, handling, and supply—75 MMSCFD	GCS—8 nos. Gas production, handling, and supply—300 MMSCFD
Workover rig	Owned—1 no. (two shifts a day operation)	Owned—1 no. (round-the-clock operation)
Well completion and testing	One team	Two teams
Cementing	Cementing unit—7 nos. Bunkers—5 nos.	Cementing unit—7 nos. Bunkers—5 nos.
Logging	1 unit (on contract)	2 units (on contract)
Subsurface	Reservoir data acquisition team—1 no. Area monitoring team—1 no.	Reservoir data acquisition team—2 nos. Area monitoring team—2 nos.
Engineering services	*Repair and maintenance* Mechanical workshop Electrical workshop *Works* New project team Civil works and maintenance	*Repair and maintenance* Mechanical workshop Electrical workshop *Works* New project team (laying of 263-km pipeline) Civil works and maintenance
Other related facilities and Support services	Mud services, Well services, HR, Finance, Supply chain management, IT and telecommunication, Logistics and auto, General management, Health services, PR (Corporate Communication), Legal, Audit, Security, Fire, and others	

10.8.5 Analysis of the Existing Manpower

The structure of the Asset is shown in Figure 10.1. It is headed by an Asset manager with the following groups/departments/teams under him: drilling, cementing, mud services, surface, subsurface, well services, logging, engineering services, logistics and auto, IT and telecommunication, HR, general management, public relations (PR), legal, audit, health services, finance, supply chain management, and few others.

Both qualitative and quantitative analyses of the current manpower inventory of the Asset reveal many interesting facts that are presented in the following paragraphs:

(a) The composition of the Asset including strength of all groups/departments/teams is shown in Figure 10.2. Drilling services lead with 34% of Asset strength (drilling—27%, cementing—3%, and mud—4%), followed by surface group—11%, logistics and auto—10%, engineering services—9%, HR—7%, and so on.

(b) The strength of officers—57%—is far more than workers—43%—in the Asset. The skilled workers account for 24% and unskilled workers 19% (Fig. 10.3). The same trend is observed throughout the organization.

Figure 10.4 shows that officer to worker ratio is around 1.33:1. In other words, there is one worker for every 1.33 officers. This is an area of concern, especially in an E&P company whose operations are predominantly field oriented. The promotion policy of the company is believed to be the reason for such skewed distribution, which needs serious introspection and corrective measures.

(c) The Asset is situated in a remote area/part of the country and lacks many basic amenities and facilities such as quality education, health services, recreation facilities,

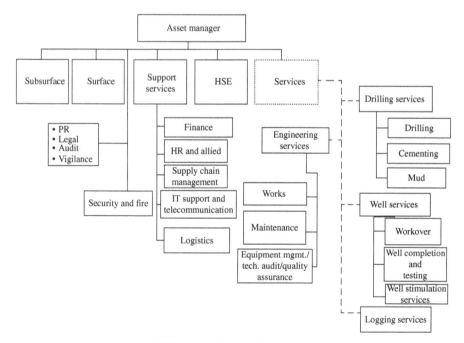

FIGURE 10.1 Asset structure.

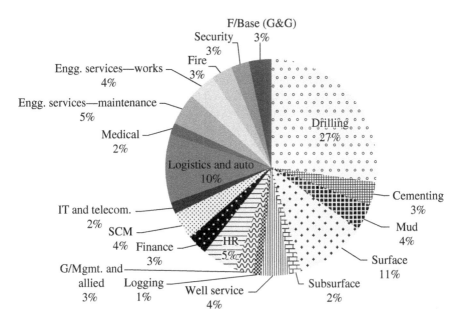

FIGURE 10.2 Group-/department-wise strength of Asset (total: 1005).

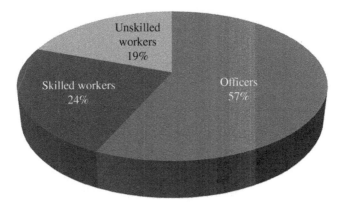

FIGURE 10.3 Officer and worker strength (total: 1005).

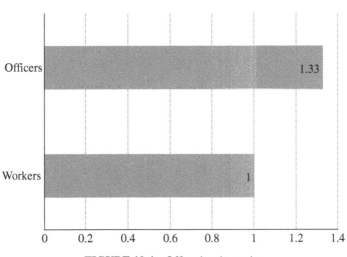

FIGURE 10.4 Officer/worker ratio.

restricted mobility, and so on. Therefore, many officers do not bring their families while posted here. They avail quarterly travel facility (QTF) as per the company policy. Mostly senior- and middle-level officers avail this facility. This affects the continuity of work and is often found as hindrance to smooth functioning of the Asset. Figure 10.5 shows that around 20% of officers avail QTF, which means one officer out of every five is always on QTF and is away from the Asset. This shows location-/Asset-specific variability, which is to be taken into account for manpower planning.

(d) The skill base/qualification of employees was also analyzed. It was observed that around 47% of officers are professional degree holders, 24% are professional diploma holders, and the remaining 29% are nonprofessional degree/diploma holders and below (Fig. 10.6). This is an area of concern and reaffirms the slack or compromised promotion policy of the organization.

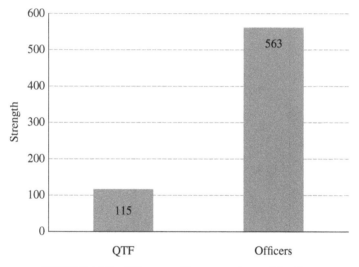

FIGURE 10.5   Officers availing quarterly travel facility.

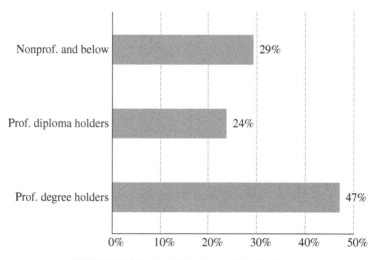

FIGURE 10.6   Qualification profile of officers.

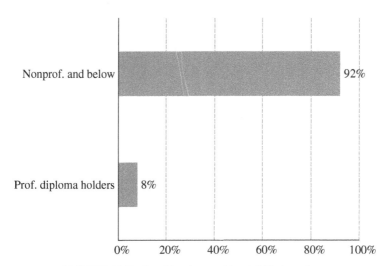

FIGURE 10.7 Qualification profile of skilled workers.

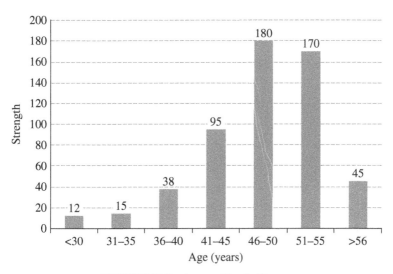

FIGURE 10.8 Age profile of officers.

Similarly, only 8% of skilled workers are professional diploma holders, and the remaining 92% are nonprofessional degree/diploma holders and below (Fig. 10.7).

(e) The age profile of officers, skilled workers, and unskilled workers is shown in Figures 10.8, 10.9, and 10.10, respectively. The average age of officers is 47.7 years, skilled workers 48.3 years, and unskilled workers 47.9 years. The employees are comparatively aged, which is an area of concern, especially the field personnel, crew, and unskilled worker whose activities are mostly labor oriented.

(f) The retirement profile of employees for the next 15 years is shown in Figure 10.11. About 133 employees are retiring in the next 5 years; thereafter, retirement rate will continue to rise for the next 8 years. This calls for appropriate induction and succession planning.

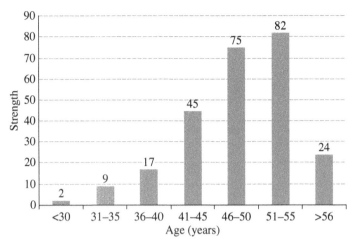

FIGURE 10.9 Age profile of skilled workers.

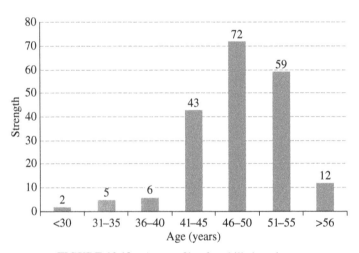

FIGURE 10.10 Age profile of unskilled workers.

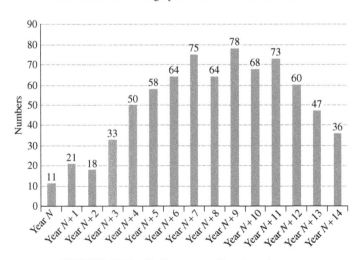

FIGURE 10.11 Retirement profile of employees.

Further breakup of retiring employees (officers, skilled workers, and unskilled workers) for the next 15 years is shown in Figure 10.12. About 71 officers, 43 skilled workers, and 19 unskilled workers are retiring in the next 5 years.

The skill base/qualification profile of retiring employees for the next 15 years is shown in Figure 10.13. Around 20 professional degree holders, 15 professional diploma holders, and 98 nonprofessional degree/diploma holders and below are separating in the next 5 years.

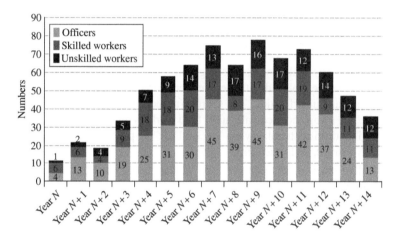

FIGURE 10.12   Retirement profile of employees—breakup.

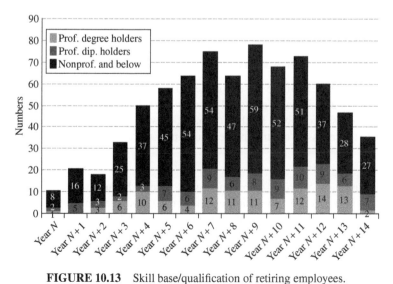

FIGURE 10.13   Skill base/qualification of retiring employees.

10.8.5.1 Suggestions: Based on the analysis and findings in Section 10.8.5, the following suggestions are made to improve the quality and suitability of personnel posted in the field, operational, functional, and other areas:

(a) In order to have appropriate skill base/qualifications of employees, especially for officers, and to have ideal officer to worker ratio, the promotion policy of the organization needs to be reviewed. The current promotion policy of natural progression of employees compromises with the skill base/qualifications required for the post. The qualification and skill base required at the induction level for officer grade should be strictly followed and should not be compromised for the in-house employees/candidates.

(b) The organization is having an aging population—the average age of employees is 47.9 years. In order to cope with laborious and strenuous field jobs that require agility and physical fitness, it is suggested that aging and medically unfit field workers may be replaced with the younger people with desired physical attributes, skill sets, and qualification. The short-term recruitment of such posts may be initiated. The aging and medically unfit field workers may be posted to less laborious jobs like drilling tubular yard stock (DTYS), production tubular yard stock (PTYS), stores, and base office.

(c) Unskilled workers account for 19% of the regular employees; there is scope for rationalizing their strength and engage them on contract barring the legal requirement.

(d) A significant number of employees are retiring in the next 5–10 years. In order to cope with this attrition, appropriate succession plan as well as induction plan may be prepared and implemented.

(e) The policy of quarterly leave fare (QTF) followed in the company needs review, or suitable allowances may be considered while assessing manpower requirement.

10.8.6 Assess Manpower Requirement

Manpower requirement of the Asset for the current (Year N) as well as future activity levels (Year $N+3$) has been assessed based on (i) the manning norms, work practices, and processes followed by the company, as well as (ii) the best practice benchmark. Both these assessed manpower demands were compared with the current strength:

(a) The summary of group-/department-wise assessed manpower requirement as per (i) the best practice benchmark and as per (ii) organization's manning norms and work practices vis-à-vis current manpower strength and corresponding surplus/deficit position is shown in Appendix 10.B.

(b) Further, activity-/subsection-wise breakup of assessed manpower demand as per (i) the best practice benchmark and (ii) organization's manning norms and work practices has been compared and is shown in Appendix 10.C (10.C.1–10.C.6).

(c) Some notable excess/deficit categories in various groups/departments as per both these methods are shown in Appendix 10.D.

(d) The illustrative working sheets detailing category-wise, activity-/subsection-wise assessed manpower requirement of various groups/departments are shown in Appendix 10.E (10.E.1–10.E.4). The assessment is based on the organization norms and work practices and is compared with the current manpower strength to arrive at the respective surplus/deficit categories/disciplines.

10.8.7 Findings and Results

10.8.7.1 Summary of Assessment: Appendix 10.B summarizes the manpower requirement of all groups/departments at the stated activity level and compares them with the current manpower to find out the surplus/deficit. It may be seen from Appendix 10.B that:

(a) The current manpower strength is sufficient for most of the groups/departments as per the assessment based on (i) the organization's manning norms and work practices and (ii) the best practice benchmark.

(b) *The current regular manpower strength is 123% higher than the assessed manpower demand based on the best practice benchmark (comparable current strength is 943 against the requirement of 423 as per the best practice benchmark—a surplus of 520 regular employees in the Asset).* On the other hand, *the current manpower strength is 10% higher than the assessed manpower demand based on the organization norms and work practices (at Year N).* This would be marginally higher at 1.4% in Year $N+3$.

(c) Similarly, the current contract employee strength is 114% higher than the best practice benchmark *(current strength is 103 against the requirement of 48 as per the best practice benchmark—a surplus of 55 contract employees in the Asset). The current contract employee strength is 20% higher than the assessed strength based on the organization norms and work practices (at Year N).*

(d) There is a huge difference between the assessment based on (i) the best practice benchmark and that of (ii) the organization's manning norms and work practices. *The assessed manpower demand as per organization norms and work practices is **120% higher** than that of the best practice benchmark (requirement as per organization norms—930—and that of best practice benchmark—423). Such huge difference is mainly because the best practice benchmark follows tighter norms, multitasking, and multidisciplinary approach, farms out noncore business/activities, emphasizes on automated production installations, engages unskilled workers on contract, and so on. Improvements in these areas are called for.*

Manpower assessment of *security and forward base (G&G)* was not done due to the following reasons:

Security: The Asset is situated in a disturbed area where security is an important issue. It is requisitioned from the state police on contract whose strength is dependent on the threat perception, duty points, and so on, and is determined by the state intelligence, police, and Asset's security department. **Forward base** is a G&G Group engaged in exploration work and is part of another Basin, although housed in this Asset for historical reasons.

The current manpower strength of the Asset is 1005 regular employees and 105 contract employees. In order to carry out like-to-like comparison between the assessed manpower demands and compare them with the available strength, these two groups/departments were excluded. *Thus, the comparable strength of 943 regular employees and 103 contract employees was considered for comparison and rationalization.*

10.8.7.2   Comparison of Assessment between Two Methods: Appendix 10.C (10.C.1–10.C.6) provides the activity-/subsection-wise breakup of the assessed manpower demand for various groups/departments and compares them following (i) the best practice benchmark and (ii) the organization's manning norms and work practices. It identifies the areas of major differences following these two assessment methods, underlying reasons for variations, and scope for bridging the gap in various groups/departments, which are given as follows:

(a) **Drilling services**: The huge difference is due to the loose manning norms followed by the organization at drilling rigs and base office. Apart from this, a large number of unskilled labors are engaged at drill sites, which may be brought to the level of the best practice benchmark.

(b) **Cementing services**: The existing unskilled workers account for more than 37% of the department strength, which may be reduced and engaged on contract.

(c) **Mud services**: The unskilled workers account for one-third of the department strength, which may be reduced and engaged on contract.

(d) **Surface**: The huge difference is due to the fact that the best practice benchmark considers automated GCS and hiring of related services. It also emphasizes on multiskilling of operational crew. The same may be adopted as the manning norms of the company.

(e) **Subsurface**: The best practice benchmark suggests broadening of scope of subsurface and strengthening of the group/team.

(f) **Well services**: The considerable difference is due to slack manning norms of workover rig, well testing, and completion team, which are quite high compared to the best practice benchmark.

(g) **Engineering services—maintenance**: The best practice benchmark suggests outsourcing of major repairs and maintenance; only fabrication, welding, and lathe operations are to be done in-house.

(h) **Engineering services—works**: Further outsourcing and restructuring of the group and tighter organization norms are needed to bridge the gap with the best practice benchmark.

(i) **IT and communication**: The current strength is 200% higher than the comparable best practice benchmark.

(j) **Logistics**: The current strength is 65% excess over the comparable best practice benchmark and comprises 25% unskilled workers. There is much scope for outsourcing.

(k) **Auto**: The current strength is in excess of 300% of the comparable best practice benchmark. There is enormous scope for outsourcing.

(l) **HR**: The current manning of core HR functions is three times higher than the comparable best practice benchmark. Unskilled workers account for 23% of the current strength. There is considerable scope for reduction both in core (personnel affairs and data, labor relations, training, and recruitment) and noncore (estate, hospitality, land acquisition, and general administration) HR functions.

(m) **Medical services**: The part-time doctors are also available on contract in dispensary and drill sites. Pharmacists may be engaged on contract.

(n) **Finance and account**: The current strength is 138% higher than the best practice benchmark, and unskilled workers account for 20% of the department strength.

(o) **Supply chain management**: The current strength is three times higher than the best practice benchmark and comprises 40% unskilled workers whose strength may be rationalized and engaged on contract.

(p) **Fire**: The current strength is 15–20% excesses over the developing country benchmark.

10.8.7.3   Identifying Notable Surplus/Deficit Categories: Appendices 10.B and 10.C do not reveal many important issues, for example, there may be shortages in some critical categories that are overshadowed by excesses in other categories within the same group/department. As a result, the genuine requirement goes unnoted or does not attract desired attention, thus affecting the efficient and smooth functioning of work. Appendix 10.D takes care of this; some notable excess/deficit categories as per (i) the organization manning norms and work practices and (ii) the best practice benchmark are shown in Appendix 10.D. It has been derived from Appendices 10.C and 10.E:

(a) It may be seen from Appendix 10.D that there are acute shortages in the category of Topman and Rigman in drilling discipline, but these are clouded due to excesses in the category of Jr.-level drilling engineer. In order to rationalize manpower strength, planners often take the recourse of leveling the excess manpower with that of deficit categories. But it should be done with extreme care taking into account many humane factors. For example, shortages in Topman can be partially met from excesses in Jr.-level drilling engineer category but not entirely, as the job of Topman and Rigman are labor intensive and require physical fitness, agility, and comparatively younger age. Most of the Jr.-level drilling engineers are 52+ years of age, risen from the rank, and not as agile and physically fit as required for the job of Topman. But, the head count shows there is sufficient manpower in the drilling group. These factors are to be considered for determining the net requirement or shortages.

(b) The excess in these categories is due to promotion policy of the organization wherein Topman and Rigman become Jr.-level drilling engineers on natural progression without enhancing their skill sets and qualification required for the post. These are true in many other categories and disciplines across the organization.

(c) The excess manpower is found in the category of "unskilled workers" in many departments, such as drilling, cementing, surface, HR, medical, finance, supply chain management, IT and telecommunication, and so on. The "unskilled workers" account for 19% of the Asset strength, and there are considerable excesses as per assessment. Further, this category is recommended to be engaged on contract barring the legal minimum.

10.8.7.4 Assessment Sheet: Illustrative   Appendix 10.E shows category-wise and activity-/subsection-wise assessed manpower requirement of various groups/departments based on the organization norms and work practices. These are illustrative working sheets that would provide a fairly good idea for carrying out the detailed assessment of manpower requirement. The assessment is compared with the current manpower strength to determine the respective surplus/deficit categories and disciplines. Appendix 10.E also reveals the reasons for slack organization manning norms and areas for improvement. Some of these norms and work practices are based on convenience rather than the actual assessment of workload. Appendix 10.E contains some of the worksheets (10.E.1–10.E.4) and not all due to space constraint. All these appendices, namely, Appendices 10.B, 10.C, 10.D, and 10.E, are illustrative only.

10.8.8   Suggestions

Based on the detailed analysis and results, the following suggestions would help in optimizing manpower requirement, improving employee utilization and productivity and aligning with the best practice benchmark:

(a) The manning norm followed by the company appears to be slack and outdated compared to the best practice benchmark. This needs to be modified across the organization, as the difference with the best practice benchmark is huge (123%). *There is much scope for improvement by discouraging compartmentalization of work, adopting multiskilling and multidisciplinary approach, outsourcing low-end services and noncore activities, automating production installations, engaging unskilled workers through contract, and so on.*

(b) Broaden the scope of operating crew at deep drilling rigs, production and process installations, and well services with suitable multiskilling and multidisciplinary orientation, which would ensure better crew utilization, less manpower, and higher productivity.

(c) Encourage automation in production and process installations and other places wherever feasible. The huge difference in manning of production installations and surface facilities between the best practice benchmark and that of organization norms is due to multiskilling crew and greater automation for the best practice benchmark, whereas prevailing manning norms emphasize on manual operations and compartmentalized work practices.

(d) Broaden the scope of geoscientists with intensive multiskilling training in geology, geophysics, reservoir, geochemistry, and so on, with a view to develop specialist cadres.

(e) Similarly, broaden the scope of maintenance crew at the deep drilling rigs, well services, and production installations. The best practice benchmark follows multiskilling maintenance crew responsible for all routine maintenance jobs at the deep drilling rigs, production installations, and well services, whereas prevailing manning norms of the organization emphasize on separate mechanical, electrical, and instrumentation maintenance crew in the respective operational areas. *Merge the role of electrical, mechanical, and instrumentation operators as "technician" for this purpose.*

(f) Strengthen the central maintenance team or field party maintenance at the base to attend relatively major breakdown and preventive maintenance that cannot be done by the routine maintenance team at the deep drilling rigs, production installations, and well services.

(g) Significant outsourcing is possible for off-site maintenance and major repairs, which are presently done in-house in the mechanical workshop, electrical workshop (repair of motors, rewinding of LT motors, and replacement of bearings), diesel shop (repair of diesel engine, top overhaul of engine, and minor engine repairs), assembly shop (rotating equipment repair, centrifugal pump repair, and SRP gearbox repair), auto workshop (repair of light vehicle, heavy vehicle, and heavy equipment), and so on. *Encourage outsourcing of off-site maintenance to the OEM, OES, or their representatives, authorized workshop, and so on, through long-term contract.*

(h) Outsource low-end, noncore business activities that are readily available in the market (e.g., housekeeping, civil works and maintenance, electrical maintenance, office and colony maintenance, instrumentation maintenance, pump house operation, engagement of unskilled workers, logging services, drilling services, shot hole drilling, etc.) and concentrate on core activities.

10.8.9 Benefits

An enterprise may strive for adopting the best practice benchmark as its manning norm. This may be done in a phased manner within a fixed time frame. This cannot be ignored for long in this competitive business world. The previous case study shows that the current regular employee strength is **123% higher** than the best practice benchmark (comparable current strength is 943 against the best practice benchmark of 423, **a surplus of 520 employees**). Considering

20% allowances for local factors, legal or statutory requirements, and so on, the requirement as per the best practice benchmark would be 507—even then **a surplus of 436 employees**.

Considering the average cost per employee per year is USD 43,515 (as per the published annual report of the company), *the savings on account of the reduction of regular manpower to the best practice benchmark with 20% allowances would be* **USD 19 million** (at USD 43,515×436 excess employees). Similarly, there would be savings on account of reduction of contract employees to the best practice benchmark.

This savings is only for one Asset, and if the best practice benchmark manning norms with 20% allowances is replicated in the entire organization, the savings would be enormous. Such drastic reduction may not be possible in few years; therefore, *considering a modest 20% reduction of the current manpower strength of the entire organization for gradual adoption of the best practice benchmark with necessary allowances, the savings would be* **USD 290 million** (at USD 43,515×20%×33,270 employees).

Apart from the savings on account of reduction/optimization of manpower, there are many indirect and hidden benefits like improved employee utilization, focused training and development, rationalization of workforce, streamlining of work, and improved morale.

10.9   CONCLUSION

Much attention in E&P companies is focused on the core activities, but there are many functional areas that involve huge expenditure and have considerable scope for improvement. The enormous scope for reduction in expenditure through manpower optimization and improving employee productivity as shown in this study is a case in point. The challenges of the upstream sector lie in managing E&P as well as functional activities *at a given cost and not at any cost.*

Manpower optimization has been carried out in this study emphasizing on multiskilling and multidisciplinary approach, rationalizing more than one hundred disciplines and subdisciplines prevailing in the organization to a manageable level at par with the industry best practice benchmark, impressing on the need for having right organization structure, identifying flaws in the existing planning process in the organization, and suggesting corrective measures. Further, a case study has been presented showing the process of manpower optimization that includes demand forecasting, supply analysis, and suggestion of means to bridge the gap.

Manpower requirement was assessed using the (i) best practice benchmark and (ii) organization's manning norms and work practices. Both these assessments were compared, and suggestions were offered to improve the organization's manning norm. Both qualitative and quantitative analyses of the existing manpower were carried out for matching the demand. The surplus and deficit categories/disciplines were identified, and useful suggestions were made to optimize manpower for greater efficiency and to align with the best practice benchmark. These suggestions are summarized in the following paragraphs for convenience:

- There is considerable scope for manpower optimization in the organization, which may be achieved through the following means: (i) aligning company's manning norms and work practices with the best practice benchmark, (ii) adopting multiskilling and multidisciplinary approach, (ii) decompartmentalization of work and broadening the scope of work in identified areas, (iii) outsourcing low-end services and noncore activities, (iv) automating production and process installations, (v) engaging unskilled workers through contract, and so on.

(i) Manning norms followed by the organization appear to be slack and outdated compared to the best practice benchmark and is mostly based on judgment rather than actual assessment of workload. The best practice benchmark with some allowances for local factors and statutory or legal requirements may be adopted as manning norms of the company, wherein local factors depend on the maturity of local market, infrastructure, degree of industrialization, and so on.

(ii) Identify disciplines and categories where multidisciplinary and multitasking may be introduced for improving utilization and broadening the areas of operation of employees, for example, maintenance crew at drilling rigs, well services, and production installations; operational and maintenance workers in cementing services; and several categories in logistics department, production installations, and many other areas, which have been elucidated in Section 10.6.2.

(iii) There is scope for broadening and decompartmentalization of work in many areas and categories, such as operating crew; maintenance crew at drilling rigs, well services, and production installations; geoscientists in various G&G disciplines; and so on (refer to Section 10.8.8).

(iv) The organization is currently having more than 100 disciplines/subdisciplines/categories owing to various reasons. Due to the advancement in technology and greater scope for outsourcing, many disciplines/categories have become redundant, which have been suggested to abolish and/or merge. It is also suggested to consolidate the large pool of disciplines/categories to a manageable level of around two dozen as per industry best practice. The illustrative list of these disciplines/categories has been provided in Section 10.7.6.

(v) Outsourcing off-site maintenance and major repairs to OEMs, OESs, low-end services, and noncore activities, which have been elaborated in details in Section 10.8.8.

(vi) Introducing automation in production and process installations and other areas (refer to Section 10.8.8).

(vii) The promotion policy of the organization needs to be reviewed that has resulted in inappropriate officer to worker ratio. The policy of natural progression of in-house employees compromises with the skill sets and qualifications needed for the post, especially for officer grade.

(viii) The organization is having an aging population (average age 47.9 years), which is affecting laborious and strenuous field jobs. The aging and medically unfit field workers may be replaced with younger people with desired physical attributes, skill sets, and qualifications. Medically unfit people may be posted in less laborious jobs such as in DTYS, PTYS, stores, offices, and so on, based on their suitability. Alternatively, they may be induced for early retirement.

(ix) Engage unskilled workers (which account 19% of regular employees) on contract barring legal requirement. Prepare succession plan to cope with the retirement of large number of experienced personnel in the next 5 years.

(The detailed suggestions are presented in Sections 10.4.1.1, 10.4.2.1, 10.6.2, 10.7.6, 10.8.5.1, and 10.8.8.)

10.9.1 Benefits

The previous suggestions would help in optimizing manpower strength of the company. There is a huge scope for improvement, but it's not an easy task to implement within a short period. Considering a modest 20% reduction of current manpower for gradual adoption of the best practice benchmark with suitable provisions for local factors and statutory or legal requirement, the savings would be USD 290 million. Besides, it would enrich the quality of manpower, streamline workforce, and improve organizational efficiency.

Chapter 11 deals with business process simplification, which is a powerful technique to improve system efficiency in an organization. It discusses the basic concept and nuances of business process simplification and business process redesign/reengineering and illustrates some real-life examples of business process simplification for enhancing organizational efficiency.

REVIEW EXERCISES

10.1 Give some examples where multiskilling and multidisciplinary approach may be introduced in your organization. What are the procedures you would follow to introduce it?

10.2 What are the procedures you would follow to identify low-utilization disciplines and categories?

10.3 What do you understand by rationalization of disciplines and categories? Is there any scope for implementing it in your organization?

10.4 What are the procedures you would follow to introduce it in your organization?

10.5 What are the reasons for having large number of disciplines and categories in the organization under study?

10.6 What are the reasons for redundancy in large number of disciplines and categories in the organization under study?

10.7 What are the factors that influence manpower demand forecasting? What are the techniques used for demand forecasting?

10.8 What are the factors considered for predicting manpower supply availability from (i) internal/in-house and (ii) external market?

10.9 Does organization structure have any bearing on manpower requirement?

10.10 As a manpower specialist, how would you optimize manpower in your organization? What are the methods generally used in manpower planning?

APPENDIX 10.A SUMMARIZED WORK SAMPLING OBSERVATION SHEET OF RIG MAINTENANCE CREW (ELECTRICAL MAINTENANCE CREW)

Crew	Electrical Engineer	Technician #1	Technician #2	Technician #3
Total observations	600	600	600	600
Worker working	—	28	28	30
Worker inspecting	30	40	30	38
Worker engaged	74	22	18	26
Worker idle	18	36	30	42
Worker away	478	474	494	464
% working	17.3	15	12.7	15.7
Allowances (%)	10	10	10	10
% working with allowances	27.3	25	22.7	25.7
Float (%)	72.7	75	77.3	74.3
σ (standard error of proportion)	0.0357	0.0357	0.0332	0.0363
Confidence level (%)	95	95	95	95

Summarized Work Sampling Observation Sheet of Rig Maintenance Crew (Mechanical Maintenance Crew)

Crew	Mechanical Engineer	Technician #1	Technician #2	Technician #3
Total observations	600	600	600	600
Worker working	15	80	84	106
Worker inspecting	42	38	12	24
Worker engaged	100	30	46	76
Worker idle	10	44	42	54
Worker away	433	408	416	340
% working	26.1	24.7	23.7	34.3
Allowances (%)	10	10	10	10
% working with allowances	36.1	34.7	33.7	44.3
Float (%)	63.9	65.3	66.3	55.7
σ (standard error of proportion)	0.0439	0.0431	0.0425	0.0474
Confidence level (%)	95	95	95	95
Accuracy (%)	5	5	5	5

$\sigma = \sqrt{\dfrac{p(1-p)}{N}}$ where p is the % working and N is the number of observations (per day or shift).

For example, $\sigma = \sqrt{\dfrac{0.15(1-0.15)}{100}} = 0.0357$.

APPENDIX 10.B ASSESSMENT OF MANPOWER DEMAND VIS-À-VIS CURRENT EMPLOYEE STRENGTH OF THE ASSET (BALANCING MANPOWER DEMAND AND SUPPLY)

Group/Department	Existing Manpower Strength		Assessed Manpower Demand as per				Surplus/Deficit				Remarks
			Company Norms and Work Practices		Best Practice Benchmark		w.r.t. Company Norms		w.r.t. Best Practice Benchmark		
	Regular	Contract	Regular	Contract	Regular	Contract	Regular	Contract	Regular	Contract	
Drilling	274	58	205	72	102	28	69	−14	172	30	The huge difference is due to manning of drilling rigs. Shortages in Topman and Rigman are overshadowed by the excesses in other categories
Cementing	35		23		14		12		21		Unskilled workers comprise more than 37% of the dept. strength
Mud	39	18	35		24	8	4		15	−8	Additional mud laborers at 19 person/rig exist; current unskilled worker strength at 33% of dept. strength
Surface	108	18	175		31	6	−67	18	77	12	Best practice benchmark considers fully automated GCS; therefore, huge gap w.r.t. orgn. norm and work practice; contract for instrument maint. and upkeeping of plant exists
Subsurface	18	1	23	4	27	6	−5	−3	−9	−5	Best practice suggests broadening of scope of subsurface and strengthening of the group/team
Well services	44		66		17		−22		27		The huge difference is due to manning of workover rig and well testing and completion team

Logging	12		12	8	0		4		Currently, logging services are on contract; best practice benchmark for own unit suggests 15 regular and 3 contract employees
G/mgmt. and allied	28		25	23	3		5		Includes Asset manager's cell, PR, audit, legal, HSE equip. mgmt./tech. audit/ quality assuran., etc.
HR	53	1	46	32	7		21	1	The current manning of core HR functions is three times higher than the comparable best practice benchmark. Unskilled workers account for 23%
Medical	17	10	13	11	4	0	6	10	Doctors are also available on contract in dispensary and drill sites; pharmacist and other paramedical staff may be taken on contract
Finance	31		24	13	7		18		Current strength is 138% higher than the best practice benchmark and comprises 20% unskilled workers
SCM	42	3	32	14	10	3	28	3	Current strength is three times higher than the best practice benchmark and comprises 40% unskilled workers
IT and telecom	17	1	18	6	-1	1	11	1	Current strength is 200% more than the comparable best practice benchmark
Logistics and auto	103	5	99	51	4	5	52	5	Logistics: current strength is 65% excess over comparable best practice benchmark; unskilled workers account for 25%. Auto: current strength is in excess of 300% of comparable best practice benchmark

(Continued)

| Group/Department | Existing Manpower Strength | | Assessed Manpower Demand as per | | | | Surplus/Deficit | | | | Remarks |
| | | | Company Norms and Work Practices | | Best Practice Benchmark | | w.r.t. Company Norms | | w.r.t. Best Practice Benchmark | | |
	Regular	Contract	Regular	Contract	Regular	Contract	Regular	Contract	Regular	Contract	
Engg. services—maintenance	48	5	43		15		5	5	33	5	Best practice benchmark suggests outsourcing of major repairs and maintenance; only fabrication, welding, and lathe operations are to be done in-house
Engg. services—works	42		61		8		−19		34		Further outsourcing, restructuring of the group, and tighter organization norms are needed
Fire	32	1	30		27		2	1	5	1	Current strength is 15–20% excesses over developing country benchmark
Comparable strength	**943**	**103**	**930**	**86**	**423**	**48**	**13**	**17**	**520**	**55**	
Security	28		28		28		0	0	0	0	Assessment of security and F/base (G&G) was not done for the reason stated in Section 10.8.7.1 (last para.), and existing strength is considered as requirement
F/base (G&G)	34	2	34	2	34	2	0	0	0	0	
Asset total	1005	105	992	88	485	50	13	17	520	55	

For Illustration Purpose

APPENDIX 10.C   ASSESSMENT OF MANPOWER DEMAND AND COMPARISON (ACTIVITY/SUBSECTION WISE)

10.C.1   Assessment of Manpower Demand: Drilling Services Group

Assessment Based on	Surplus/ Deficit	Existing Manpower	Assessed Manpower Demand	Distribution of Assessed Manpower Demand									Remarks
				D/Rigs (3 Nos.)	Charter Rig (2 Nos.)	F/Maint. Party	Fishing	Rig Movement	HSE and CMT	Plan. and Contract	DTY and Supply	Office Support	
Organization norms and work	69	274	205	162	6	13	1	1	4	6	5	7	The huge difference is due to manning of drilling rigs
Best practice benchmark	172	274	102	86	4	4	1	1	2	1	2	5	

Contract persons: best practice benchmark—28; orgn. norms and practice—72.

10.C.2   Assessment of Manpower Demand: Cementing Services

Assessment Based on	Surplus/ Deficit	Existing Manpower	Assessed Manpower Demand	Distribution of Assessed Manpower Demand				Remarks
				Operation	Maintenance	Plan. and Contract	Tech. and Base Support	
Organization norms and work practice	12	35	23	14	4	1	4	
Best practice benchmark	21	35	14	8	3	1	2	

10.C.3 Assessment of Manpower Demand: Mud Services

| Assessment Based on | Surplus/ Deficit | Existing Manpower | Assessed Manpower | Distribution of Assessed Manpower | | | | | | Remarks |
				Drilling Rigs (3 Nos.)	W/O Rig (1 No.)	Rig Coord.	Plan. and Contract	Three Labs (Oil and Water, Gas, Cement)	Tech. and Base Support	
Organization norms and work practice	4	39	35	12	4	2	2	12	3	
Best practice benchmark	15	39	24	12	5	1	1	3	2	

For Illustration Purpose

Contact persons: best practice benchmark—8; orgn. norms and practice—to be met from 72 unskilled workers under the drilling group.

10.C.4 Assessment of Manpower Demand: Surface Group

| Assessment Based on | Surplus/ Deficit | Existing Manpower | Assessed Manpower | Distribution of Assessed Manpower | | | | | | | | | Remarks |
				GCS (8 Nos.)	HSE	Plan. and Contract	Maint. Team	Colony Gas Metrg. Stn.	SCADA	Mktg. and Office Support	Chemistry	Specialist Pool	
Organization norms and work practice	−67	108	175	136	1	4	20	1	4	9			Best practice benchmark considers fully automated GCS, so huge gap w.r.t. orgn. norm and work practice
Best practice benchmark	68	108	40	16	1	2	9		4	5	2	1	

For Illustration Purpose

Contract persons: best practice benchmark—6; orgn. norms and practice—18.

10.C.5 Assessment of Manpower Demand: Subsurface Group

Assessment Based on	Surplus/ Deficit	Existing Manpower Strength	Assessed Manpower Demand	Distribution of Assessed Manpower				
				Area Team	Data Acq. Grp.	Database	Base Support	Remarks
Organization norms and work practice	−5	18	23	10	8	2	3	
Best practice benchmark	−9	18	27	11	10	3	3	

Contract persons: best practice benchmark—6; orgn. norms and practice—4.

10.C.6 Assessment of Manpower Demand: Well Services

Assessment Based on	Surplus/ Deficit	Existing Manpower Strength	Assessed Manpower Demand	Distribution of Assessed Manpower						
				WOR-1 Own Rig	WOR-1 Hired Rig	Well Testg. and Compl. (Two Teams)	Tech. Cell/ Contract	WSS/ WL/ PRYS	Base Support	Remarks
Organization norms and work practice	−22	44	66	35	2	21	2	3	3	The huge difference due to manning of W/O rig and well testg. team
Best practice benchmark	27	44	17	7		5		1	4	

APPENDIX 10.D   SOME NOTABLE EXCESS/DEFICIT CATEGORIES AS PER ASSESSMENT

Group/Department	Assessment as per Organization Norms and Work Practices		Assessment and Remarks w.r.t. Best Practice Benchmark
	Surplus	Deficit	
Drilling	Jr.-level drilling engineer—56 (mostly aged >52 years and having nonprofessional qualification or below)	Topman—24	Excesses are observed in all categories, except Topman and Rigman. The excesses are more than double compared to the organization norms. The shortages of Topman and Rigman are clouded by the excess manpower in other categories
	Mid-/Jr.-level mechanical maintenance engineer—10	Rigman—19	
	Unskilled worker—45	(Existing seven Rigman are unfit for rig operations hence effective requirement—26)	
Cementing mud	Unskilled worker—19	—	Considerable excesses in all categories. Unskilled workers may be engaged on contract
Surface (4 GCS—current)	Mid-/Jr.-level production engineer—10	Technician (Prod.)—1	Huge excesses in all categories as best practice benchmark suggests automated GCS, and low-end services may be outsourced
	Unskilled worker—6	Technician (Diesel/Fitting)—7	
Surface (8 GCS)	Unskilled worker—6	Technician (Elect.)—9	
		Mid-/Jr.-level production engineer—10	
		Technician (Prod.)—17	
		Technician (Diesel/Fitting)—23	
		Technician (Elect.)—27	
Subsurface	—	Jr.-level mechanical engineer—1	Marginal shortages observed. The best practice benchmark suggests broadening of scope and strengthening of the group
		Technician (Mech.)—2	
Well services	—	Technician (P)—14	Considerable excesses. Review manning of workover rig, well testing, and completion team. Unskilled workers may be engaged on contract
		Rigman—7	
		Technician (Elect.)—1	
Logging	—	Jr.-level geophysicist—4	Restructuring of the group/team is suggested

For Illustration Purpose

Function			
Engineering services—maintenance	Jr.-level mechanical engineer—9 Jr.-level electrical engineer—4	Technician (fitting/welding/machinist)—7	Considerable excesses in all categories, as the best practice benchmark suggests outsourcing of major repairs and maintenance; only fabrication, welding, and lathe operations are to be done in-house
Engineering services—works	Draftsman/Survey/Mason/Plumbing—6	Mid-/Jr.-level pipeline/C&M engineer—5 Sr.-level civil engineer—2 Mid-/Jr.-level civil engineer—26 Mid-/Jr.-level electrical engineer—7	Considerable excesses. There is scope of further outsourcing of noncore, low-end services Restructuring of the group and tighter organization norms are needed
IT and telecom	Unskilled worker—3 Others—1	Mid-level telecom engineer—2	The current strength is 200% higher than the comparable best practice benchmark Considerable scope for outsourcing
Logistics and auto	Operator (crane)—12 Jr.-level transport officer—2	Technician (elex./telecom)—3 Slinger—7	*Logistics:* The current strength is 65% excess over the comparable best practice benchmark *Auto:* The current strength is in excess of 300% of the comparable best practice benchmark
HR	Unskilled worker—7	—	The current manning of core HR functions is three times higher than the comparable best practice benchmark. There is considerable scope for reduction both in core and noncore HR functions
Medical	Unskilled worker—4	—	Part-time doctors (for drill sites and dispensary) and pharmacists may be engaged on contract
Finance	Unskilled worker—3 Others—1	—	Considerable excesses. Unskilled workers may be taken on contract
SCM	Unskilled worker—7	Jr.-level procurement and material officer—2 Storekeeper—10	Considerable excesses. Unskilled workers may be taken on contract
Fire	—	Fire staff—4	The current strength is 15–20% excesses over the developing country benchmark

For Illustration Purpose

APPENDIX 10.E   ASSESSMENT OF MANPOWER DEMAND USING COMPANY NORMS AND WORK PRACTICES

10.E.1   Assessment of Manpower Demand: Drilling Services

Sl. No.	Desig./Role	Surplus/ Deficit	Existing Manpower Strength	Assessed Manpower Demand	Distribution of Assessed Manpower Demand									Remarks
					D/Rigs (3 Nos.)	Hired Rig (2 Nos.)	F/Maint. Party	Fishing	Rig Movement	HSE and CMT	Plng. and Provsg.	DTY and Supply	Office Support	
1	Head drilling services and Sr.-level drilling engineer	1	3	2									2	
2	Mid-level drilling engr.	2	39	37	24	5		1	1	2	2	1	1	Manning of rigs: at RIG-1, DIC/ADIC-2, SIC-4/ ASIC-4, Tech. Suptd-1/rig
3	Jr.-level drilling engr.	55	71	16	12	1				2		1		
4	Topman (drilling)	−24		24	24									
5	Rigman	−19	17	36	36									
6	Mid-level mech. engr.	8	16	8	6		1				1			At 2 per rig
7	Jr.-level mech. engr.	9	24	15	12		2				1			
8	Technician (diesel/ fitting)	−6	8	14	12		2							
9	Mid-level elec. engr.	9	17	8	6		1				1			
10	Jr.-level elec. engr.	4	13	9	6		2				1			
11	Technician (elec.)	−12	2	14	12		2							
12	Mid-level instr. engr.	0	1	1	6[a]		1							
13	Technician (instrumentation)	0	1	1			1							
14	Jr.-level HR exec.	−2		2									2	
15	PS	1	2	1									1	[a]Preferably on contract

For Illustration Purpose

#	Designation											
16	Jr.-level engr. (auto)	1	1									
17	Welder[b]	-5	2	7	6		1					
18	Storekeeper	-3	3	6	6							
19	Unskilled worker—field	45	49	4	72[c]						3	1
20	Operator (HV)	2	2									
21	Wireless operator	2	2									
22	Tech. assistant	1	1									
	Total	**69**	**274**	**205**	**162**	6	13	1	1	4	6	5	7
	Contract	-14	58	72	72							

For Illustration Purpose

Jr. level: <8 years experience, middle level: 8–20 years experience, Sr. level: 20–28 years experience, and top management: >28 years experience.

1. Three own drilling rigs operating on 28 days on/off pattern considered.
2. Two hired rigs have been considered for operation.
3. Job contract for four welders at rigs exists and that of unskilled mud laborers at 19 persons at each site operating.

[b] Contract exists

[c] To be taken on contract

10.E.2 Assessment of Manpower Demand: Cementing Services

Sl. No.	Desig./Role	Surplus/ Deficit	Existing Manpower Strength	Assessed Manpower Demand	Distribution of Assessed Manpower Demand				Remarks
					Operation	Maintenance	Planning and Provisioning	Technical and Base	
1	Sr.-level cementing engr.	0	1	1				1	Office support may be taken from drilling services
2	Mid-level cementing engr.[a]	−1	3	4	1		1	2	[a]Supervision of hired C mtg. job
3	Jr.-level cementing engr.	−1	3	4	3	1			Operation and maintenance crew are interchangeable to each other whenever required.
4	Technician/operator (omtg.)	4	10	6	5	1			
5	Technician (elec.)	1	1	1		1			
6	Welder	1	2	1		1			
7	Technician (auto)	1	2	1		1			
8	Unskilled worker	7	13	6	5		1		
	Total	**12**	**35**	**23**	**14**	**4**	**1**	**4**	

For Illustration Purpose

Jr. level: <8 years experience, middle level: 8–20 years experience, Sr. level: 20–28 years experience, and top management: >28 years experience.

1. Additional helper/laborer required during job may be taken from the drilling rig (roustabout/laborers).
2. Presently seven cementing units and five bunkers are available.
3. On an average three to four cementing units and two bunkers are used during a cementing job.
4. Simultaneous cementing jobs in two locations are not done; one job is executed at a time.
5. Cementing jobs done in a year—17 primary, 17 secondary, and 34 allied jobs.
6. More than 37% are unskilled workers in cementing services.

10.E.3 Assessment of Manpower Demand: Mud Services

Sl. No.	Desig./Role	Surplus/ Deficit	Existing Manpower Demand	Assessed Manpower Demand	Distribution of Assessed Manpower Demand						Remarks
					Drilling Rigs (3 Nos.)	W/O Rig (1 No.)	Rig Coord.	Planning and Provisioning	Two Labs (Oil and Water, Gas, Cement)	Tech. and Base Support	
1	Mid-level chemist	−5	10	15	6		2	2	3	2	Including in-charge mud services
2	Jr.-level chemist[a]	−8	3	11	6	2			3		[a]Leave reserves for rig to be met from labs/base
3	Tech. assistant (chem.)[b]	7	12	5		2			3		[b]Two tech. astt. (chem.) in well services accounted here
4	Unskilled worker—field[c]	9	13	4					3		[c]To be taken from rigs
5	Jr.-level HR exec.	1	1							1	HR help to be taken from drilling services
	Total	**4**	**39**	**35**	12	4	2	2	12	3	

Jr. level: <8 years experience, middle level: 8–20 years experience, Sr. level: 20–28 years experience, and top management: >28 years experience.

1. Job contract for unskilled mud laborers at 19 persons exists at each drilling rig.
2. Unskilled workers account from 1/3 strength of mud services.
3. Excess unskilled staff are to be posted in logistics.

10.E.4 Assessment of Manpower Demand: Surface Group

Sl. No.	Desig./Role	Surplus/ Deficit	Existing Manpower Strength	Assessed Manpower Demand	GCS (8 Nos.)	HSE	Planning and Provisioning	Maintenance Team	Colony Gas Metrg. Stn.	SCADA	Marketing and Office Support	Remarks
1	Head surface group and Sr.-level production engineer	0	2	2							3	
2	Mid-level production engr.[a]	−10	21	31	24	1	1			2	3	Manning of installation: [a]inst. managers—1, SICs—4, shift operators—4 per installation
3	Jr.-level production engr.	0	19	19	16		1			2		
4	Technician (prod.)	−17	16	33	32				1			
5	Mid-level mech. engr.	−1	5	6			1	5				For better utilization, maintenance team of surface group will also look after the maintenance job of well services. It has been strengthened accordingly
6	Jr.-level mech. engr.	9	9									
7	Technician (diesel/fitting)	**−32**	**3**	35	32			3				
8	Mid-level elec. engr.	**−3**	2	5			1	4				
9	Jr.-level elec. engr.	6	6									
10	Technician (elec.)	−30	4	34	32			2				

For Illustration Purpose

No.	Category						Service contract exists[a]
11	Mid-level instn. engr.	0	2			2	
12	Jr.-level instn. engr.	1 ⎱	0			4	4
13	Technician (instn.)	−4 ⎰	4				[a]
14	Jr.-level HR exec.	0	1				1
15	PS	1	2				1
16	Unskilled worker—office	7	8				1
17	Jr.-level drilling engr.	7	7	0			1
	Total	**−67**	**108**	**175**	**136**	**20**	**9**
	Contract	**18**	**18**	**1**	**1**	**4**	**9**

Jr. level: <8 years experience, middle level: 8–20 years experience, Sr. level: 20–28 years experience, and top management: >28 years experience.

1. Job/service contract exists for instrumentation maintenance—6 persons; upkeeping of installations—6 persons; O&M contract for pump house—6 persons.

2. *Planning and provisioning:*
Average no. of transactions (PR) in a year: operation—14; mech.—11; elect.—11; instt.—3; value of total transaction/year—USD 0.5 million.
Planning and provisioning of mechanical and electrical items shall be continued as is done by the field maintenance party.

3. *Field maintenance party:*
Mechanical—av. preventive maintenance = 60/year; breakdown maint. = 6/year; WOR—av. prevent. maint. = 36/year; b/down maint. = 2–3/year.
Electrical—mostly lighting and electrification jobs, minor repairs like fuse failure, switch problem, and so on, are done by the shift crew; av. prevent. maint. = 125/year; b/down maint. = 200/year.

4. *Gas metering station*—Main jobs: monitoring of pressure, attending colony complaint approximately 6 per month, mainly checking of valve, nozzle, gas leakage, draining of separator, and so on. Presently, these are attended by a semiskilled staff.

FURTHER READING

Aswathappa, K., Human Resource and Personnel Management: Text and Cases, Tata McGraw-Hill Publishing Company Ltd., New Delhi, 2002.

Bartholomew, D.J., Forbes, A.F., and McClean, S.I., Statistical Techniques for Manpower Planning, 2nd edition, John Wiley & Sons, Ltd, Chichester, 1991.

Barton, J. and Gold, J., Human Resource Management—Theory and Practice, 4th edition, Palgrave Macmillan, New York, 2007.

Bennison, M. and Casson, J., Manpower Planning Handbook, McGraw-Hill Book Company, London, 1984.

Bulla, D.N. and Scott, P.M., Manpower Requirements Forecasting: A Case Example, in Human Resource Forecasting and Modelling, ed. D. Ward, T.P. Bechet, and R. Tripp, Human Resource Planning Society, New York, 1994.

Burton, W.W., Forecasting Manpower Needs—A Tested Formula, in Long-Range Planning for Management, revised edition, Harper and Row, New York, 1958.

Charles, P.G., Strategic Human Resource Manager: A General Managerial Approach, 2nd edition, Prentice-Hall, Upper Saddle River, New Jersey, 2nd edition, 2000.

Dessler, G., Fundamentals of Human Resource Management, 4th edition, Prentice-Hall, Upper Saddle River, New Jersey, 4th edition, 2015.

Grinold, R.C. and Marshall, K.T., Manpower Planning Models, Elsevier North-Holland, New York, 1977.

Jaffe, P.J. and Stewart, C.D., Manpower Resources and Utilization Principles of Working Force Analysis, John Wiley & Sons, Inc., New York, 1951.

Kleiman, L.S., Human Resource Management, Biztantra Innovations in Management, New Delhi, 2003.

Monappa, A., Managing Human Resources, MacMillan, New Delhi, 1997.

National Manpower Council, Manpower Policies for a Democratic Society, Columbia University Press, New York, 1965.

Rao, V.S.P., Human Resource Management: Text and Cases, Excel Books, New Delhi, 2002.

Snell, S.A., Morris, S.S., and Bohlander, G.W., 17th edition, Managing Human Resources, South-Western College Pub, Cincinnati, OH, 2015.

Trowbridge, C.L., Manpower Projections: Some Conceptual Problems and Research Needs, Harvard Business Review, Volume 44, 1966, pp. 166–126.

Vetter, E., How to Forecast Your Manpower Needs, Nation's Business, Volume 52, 1969, pp. 102–110.

Useful Links

Ackoff, R.L., Aronofsky, J.S., Delesie, L., and Alvarado, G.S.M., A Bibliography on Corporate Planning, available at: http://dcsh.xoc.uam.mx/curric/sergio_monroy/pdf/a_bibliog_on_corp_manpower_planning.pdf (accessed on March 8, 2016).

Lesson: 7, Human Resource Planning: Process, Methods, and Techniques, available at: http://www.psnacet.edu.in/courses/MBA/HRM/3.pdf (accessed on March 8, 2016).

11

IMPROVING ORGANIZATIONAL EFFICIENCY THROUGH BUSINESS PROCESS SIMPLIFICATION

11.1 INTRODUCTION

In previous chapters, we have dealt with core operational activities and key functional areas. In this chapter, we will deal with business process simplification, which is a powerful means to improve the efficiency of the system by eliminating duplication of work, optimizing resource requirement, and reducing process cycle time. A good and professionally managed organization is generally known for its efficient and responsive systems, procedures, and practices.

Systems and procedures are much like infrastructure that affects the efficiency and performance of an organization. They are so intricately ingrained in the structure of an enterprise that opportunity for their improvement is often overlooked. The need for quick response by reducing cycle time for various business processes and services is critical for survival and attaining excellence in today's fast-moving business world.

The business processes that are ideal candidates for process redesign or simplification for organizational improvement are supply chain management including procurement, contract, materials management, and others; finance and accounts that contain accounts payable and recoverable, capital expenditure, and administration, and so on; and human resources (HR) that cover recruitment, training and development, manpower planning, personal claims and settlement, and so on.

Generally, state-owned companies such as national oil companies (NOCs) inherit systems and procedures from the government, which may not be in sync with the present-day business requirement. Many such business processes are lengthy, cumbersome, and rigid, which breed inefficiency and result in unproductive utilization of resources. The study shows that many prevalent processes and rules in state-owned companies are archaic and

Optimization and Business Improvement Studies in Upstream Oil and Gas Industry, First Edition.
Sanjib Chowdhury.
© 2016 John Wiley & Sons, Inc. Published 2016 by John Wiley & Sons, Inc.

nonexistent in efficient private, multinational, and international (oil) companies, for example, HR business processes for drawing housing loan, permission for visiting abroad (even personal trip), obtaining "no objection certificate" for passport, furnishing surety bond for drawing petty advances/loans, permission for movable/immovable property transaction, and so on. Such government legacy may be rationalized for enhancing organizational efficiency.

The need for change in business processes ideally arises from the organization itself. The employees and stakeholders of a company are in a better position to identify areas that require improvement. Accordingly, management may prioritize and initiate process simplification exercise. Real-life process simplification examples illustrated in this chapter highlight the essence of business process simplification.

11.2 OBJECTIVES

Systems and procedures followed in many companies are archaic, which need to be modified according to the changes in business environment and making use of enabling technology. There is a need to simplify systems and processes and improve functioning of the organization in terms of faster response, better services, reduced processing time, and optimal resource requirement. Toward this end, the objectives of this study are as follows:

(a) To identify opportunities for business process simplification and improve organizational efficiency
(b) To simplify various processes to
  • Improve quality of services
  • Reduce processing/cycle time
  • Eliminate redundant and nonvalue-adding activities/steps
  • Optimize resource requirement by freeing up scarce resources, especially manpower

Pursuant to these objectives, a few prevailing business processes in a large E&P company have been simplified and illustrated in this work.

11.2.1 Methodology

The methodology adopted for the current study is based on rational approach and common sense aided by experience in the system. This is similar to "lean methodology," which has been explained in Section 11.3. The steps followed in this process simplification exercise are as follows:

  • Study the existing system and business processes in the organization
  • Map the current business processes and draw workflow diagrams
  • Identify redundant and nonvalue-added tasks/steps
  • Map the utopian state, that is, ultimate or ideal state one aspires for
  • Map the applicable state, which is feasible at the current environment

- Compare the simplified process with the best practice approach, wherever available or applicable
- Simplify further to ensure alignment with the best practice approach

The business processes are enterprise specific and vary from company to company depending on the internal and external business environment. Sometimes, a business process needs to accommodate and comply with specific government rules and regulations, which may be hindrance to align with the best practice approach.

11.3 DIFFERENCES BETWEEN BUSINESS PROCESS SIMPLIFICATION AND BUSINESS PROCESS REDESIGN/REENGINEERING

There are fundamental differences between business process simplification and business process redesign/reengineering, which are indistinguishable to many. The following paragraphs explain their basic nuances.

Business process simplification *deals with identifying and removing unnecessary activities or steps to shorten cycle time, improve customer satisfaction, and save time and resources of the organization. It is essentially a process-led evolutionary approach aimed at incremental change or improvement within the existing framework (that may require minor changes).* It is usually done as a "reactive" measure to improve the existing processes, which are considered as inefficient and unfriendly to customers. The in-depth analysis of the current process helps in identifying the pain points, which are dealt with to improve system efficiency. In short, business process simplification is an inexpensive but mighty tool to improve customer satisfaction and process efficiency. It is easy to implement and enhances the scope for application of enabling technology. It is also aided by the use of enabling technology.

The key point in business process simplification exercise is to identify the nonvalue-added activities and eliminate them. Incidentally, this is akin to *lean methodology*—a systematic method to eliminate nonvalue-added activities or Muda (waste). In lean methodology, there are seven types of Muda (wastes), namely, defects, overproduction, inventory, overprocessing, unnecessary human motion, unnecessary transport and handling, and waiting, which are normally caused by Mura (unevenness) and Muri (overdoing).

The following two types of activities are important for business process simplification:

(i) Activities that do not add value but deemed necessary for compliance of rules and statutory requirement.
(ii) Activities that do not add value and are unnecessary. These are required to be eliminated forthwith.

Therefore, lean methodology is no different than eliminating unnecessary activities (waste), reducing process cycle time, and improving process efficiency. Thus, lean approach is useful to process simplification. However, process simplification exercise predominantly follows a rational approach aided by experience and common sense.

On the other hand, **business process redesign** *popularly known as* **business process reengineering (BPR)** *is associated with fundamental rethinking and radical redesigning* *in key business processes. Contrary to business process simplification, it ignores organi-zational boundaries and aims at drastic improvement in critical performance measures such as cost, quality, service, and speed* [1–3]. BPR took the business world by storm and became a buzzword in the 1990s with the pioneering work of Michael Hammer, James Champy, Thomas Davenport, James Short, and others (see Bibliography). But it did not come entirely of its own; many factors influenced its advent, which are mentioned in the following paragraphs [4]:

- Recession and globalization in the late 1980s and the early 1990s
- Competitive business and increasing customer demand
- Advancement in computing, information technology, and IT-enabled services (ITeS)

In addition, total quality management (TQM) and Kaizen, just-in-time (JIT), and flex-ible manufacturing system (FMS) were in vogue for quite some time. All these brought business process redesign into focus of management for improving organization's performance in critical areas such as cost, quality, productivity, and customer satisfaction.

TQM and Kaizen, JIT, and FMS originated in Japanese manufacturing and logistics industry and were known to engineers and business professionals since the 1950s. *It resulted in business process improvement and became popular in the 1970s and 1980s. It signified process redesign to a certain extent. TQM is based on the concept of contin-uous and incremental improvement with emphasis on quality, effort, employee involve-ment, and willingness. It is a people-oriented approach requiring long-term discipline and is somewhat easier to implement.* Later on, computer-integrated manufacturing (CIM) and computer-integrated logistics (CIL) became successful examples of such type of business process redesign.

Business process redesign/reengineering is a customer-led approach that relies on enabling technology to improve process efficiency. It generally faces resistance from employees and is somewhat difficult to implement. Although there are success stories on the implementa-tion of BPR in iconic corporate companies (e.g., Ford Motor, Mutual Benefit Life, Capital Holding Co., Taco Bell, Walmart, Hallmark, IBM, AT&T, Sony, General Electric, Citibank, Bank of America, and others), a whopping 50–70% reengineering projects are estimated as failures. The reasons for failure are usually the following [4]:

   (i) Lack of focus, motivation, guidance, and support from the senior management
  (ii) Fixing mentality instead of changing; complacency on marginal gain
 (iii) Poor understanding of organization's need in terms of infrastructure, technology gap, capability, and lack of homework
 (iv) Inability to garner employee support or buy-in

TQM and BPR share a cross-functional orientation as per Davenport et al. [5, 6]. TQM focuses on incremental change and continuous improvement of processes over an open-ended period of time, while BPR emphasizes on sweeping change and discrete and dramatic improvement of processes within a specific time frame.

Business process reengineering has a negative connotation due to its outcome of downsiz-ing, which was not its intended purpose or primary goal. But many organizations pursued this goal vigorously, which caused anguish even to Michael Hammer—the proponent and leading founder of business process reengineering.

For further readings on business process redesign and reengineering, the interested readers may refer to "References/Further Reading" listed at the end of this chapter.

11.4 IDENTIFY OPPORTUNITIES FOR BUSINESS PROCESS SIMPLIFICATION

In order to identify opportunities for business process simplification in an organization, the following approaches may be adopted:

(a) **Benchmark current process with the best practice approach**: This would reveal the scope for improvement and gap between the current process and the best practice, which would help in adopting and aligning with the best practice approach.

(b) **Employee satisfaction survey**: Internal survey may be carried out to know the employee satisfaction level about various processes and services and inconvenience faced in the current systems and procedures. The survey may be conducted online, which is faster and easy to compile. It may also be done through circulating questionnaire among employees, if employee reach to computer is inadequate. Based on the survey result, the processes may be prioritized and process simplification may be initiated. One of the areas where business process simplification has high impact on employee satisfaction level is HR processes, especially personal claims, loan and advance, and service-related matters.

11.5 BUSINESS PROCESS SIMPLIFICATION: SOME REAL-LIFE EXAMPLES

An employee survey was conducted in a large organization operating in various locations/countries to know the pain points affecting it and its stakeholders. The internal survey revealed that employee dissatisfaction with various HR processes is very high and topped the chart. This also matched with the perception of senior management who was inundated with complaints from employees about the poor quality of HR services. In fact, there was lot of discontent among the employees over various HR services, which are lengthy, bureaucratic, and time consuming.

Therefore, HR processes were chosen as the priority area for process simplification in the organization under reference. The HR processes that needed immediate attention and can be easily implemented were identified and prioritized for process simplification. Some of these HR processes, which were simplified, are illustrated in the following subsections 11.5.1–11.5.4 as real-life examples. These would help in understanding the nuances and intricacies of process simplification approach.

11.5.1 Housing Loan

Housing is a basic human need. It is considered as a priority area to tide over the growing demand for housing in many countries and is usually encouraged by the state. The company under reference offers soft housing loan to its employees and allocates a certain amount of fund every year for this purpose. The employees are interested in availing this loan due to its low interest rate (one-third of the market rate) and easy repayment terms, which is stretched over 20 years.

The housing loan is extended to employees who have completed certain years of service (e.g., 5 years) in the company for the following purpose:

(i) Purchase of land and construction of house

(ii) Construction of house on own land

(iii) Purchase of ready-built or under-construction flat

(iv) Extension of existing house

11.5.1.1 Existing Process   The employees face lot of inconveniences due to lengthy and cumbersome processes, complex rules and regulations, unawareness about voluminous documentations required, steps to be followed, and so on.

The workflow diagram of the existing processes for the aforementioned four cases of housing loan has been mapped and is shown in Figures 11.1 and 11.2. The nonvalue-adding steps have been identified and shown in the workflow diagram (marked as shaded area). The current process for drawing housing loan for purchase of land and construction of house contains 61 steps, construction of house on own land—35 steps, purchase of ready-built flat—25 steps, and purchase of under-construction flat—39 steps.

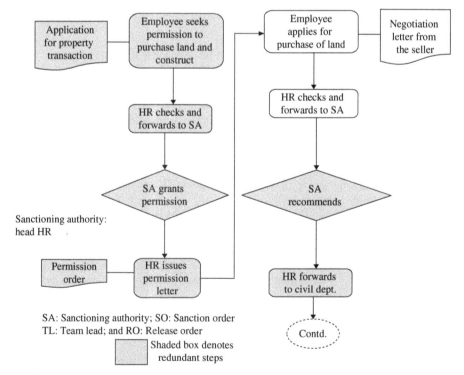

FIGURE 11.1   Loan for purchase of land and construction of house: current process.

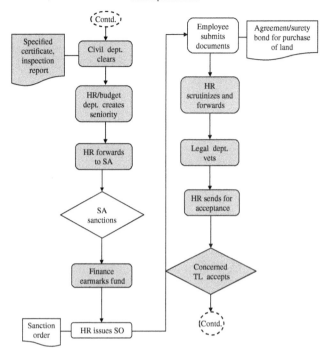

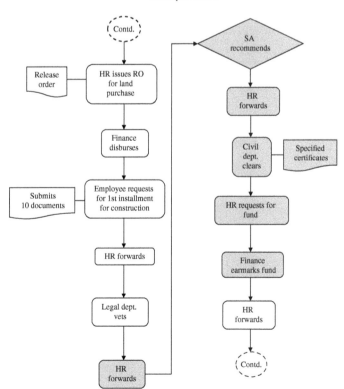

FIGURE 11.1 (*Continued*)

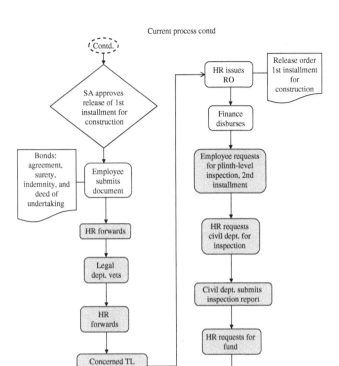

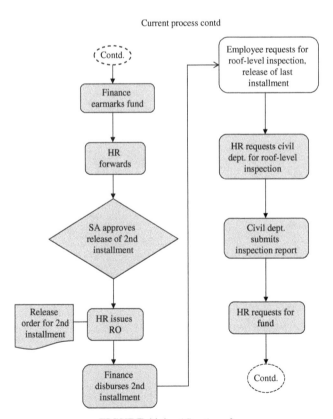

FIGURE 11.1 (*Continued*)

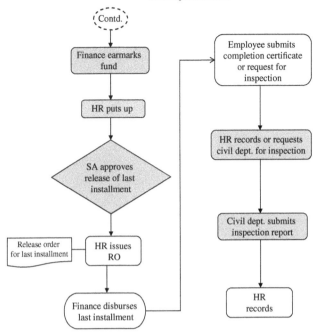

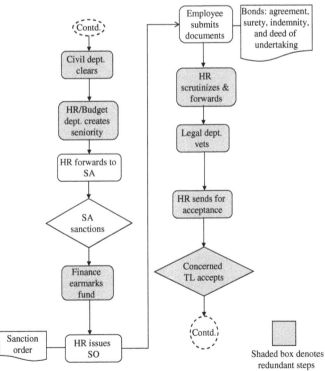

SA: Sanctioning authority; SO: sanction order; and TL: team lead

FIGURE 11.1 *(Continued)*

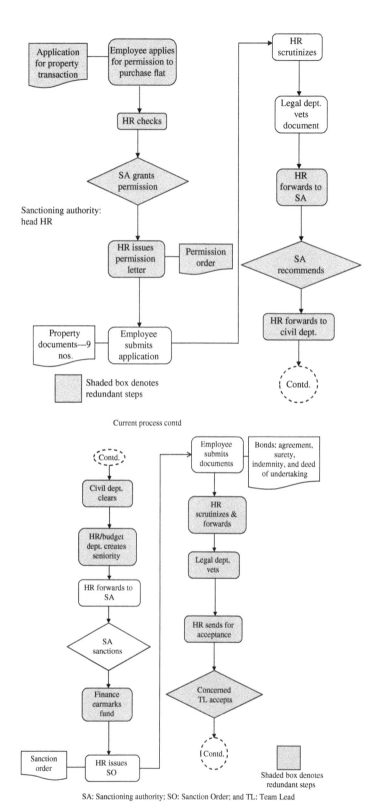

FIGURE 11.2 Loan for ready-built/under-construction house: current process.

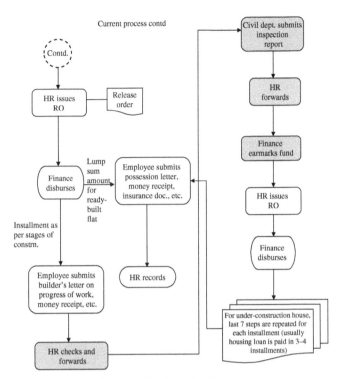

FIGURE 11.2 (*Continued*)

11.5.1.2 Simplified Process: The simplified processes for housing loan of the aforementioned cases are shown in Figures 11.3 and 11.4. It may be seen from the work-flow diagrams that

(a) The current process for obtaining housing loan for the purchase of land and construction of house *contains 61 steps, which can be simplified to 20 steps* with removal of nonvalue-added activities and procedures.

(b) Similarly, the current processes involving *35 steps for construction of house in own land, 25 steps for the purchase of ready-built flat, and 39 steps* for the purchase of under-construction flat can be *simplified to 13-step, 11-step, and 17-step process, respectively.*

(c) The employees are required to submit a number of documents, which may be further rationalized and simplified. For example, instead of submitting separate "indemnity bond," "deed of undertaking," and "agreement bond," the contents of "indemnity bond" and "deed of undertaking" may be incorporated in the "agreement bond." A unified "agreement bond" may be legally vetted and standardized; thus submission of separate "indemnity bond" and "deed of undertaking" may be dispensed with.

(d) By standardizing legal documents such as agreement bond, surety bond, indemnity bond, and deed of undertaking, only "title clearance" document may be subjected to legal vetting. Other documents as mentioned earlier are in standard legally vet-ted format; therefore, these can be verified by the concerned HR official.

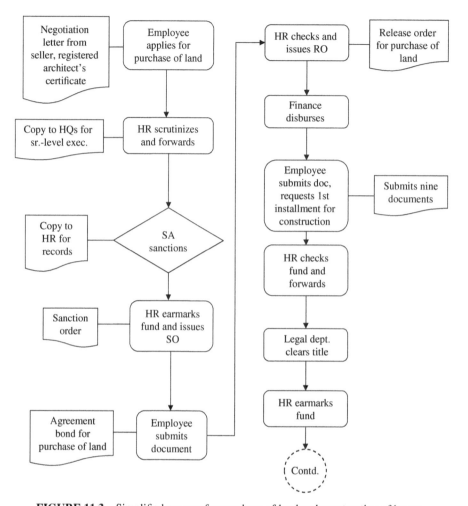

FIGURE 11.3   Simplified process for purchase of land and construction of house.

(e) The practice of inspection at each stage of construction by company officials may be discontinued. Instead, construction progress certificate from a registered architect may be submitted by employee for acceptance.

(f) In order to have adequate audit trail,

   (i) The fund may be disbursed upon receiving necessary certificate(s) from a registered architect.

   (ii) Employee must submit "completion certificate" issued by municipality or house tax bill.

(g) The aforementioned housing loan process can be further simplified by

   (i) Disbursing the loan amount in one or two installment(s), instead of multiple installments

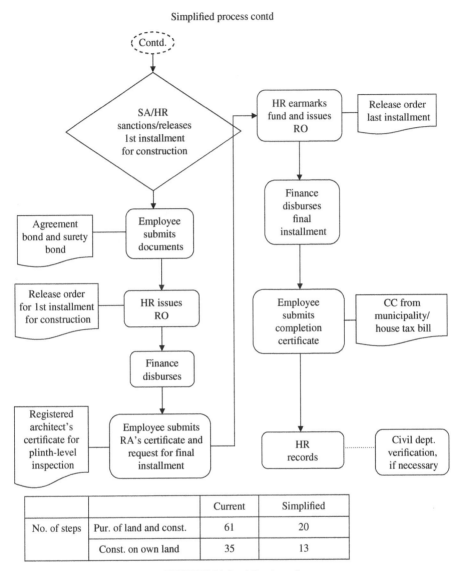

		Current	Simplified
No. of steps	Pur. of land and const.	61	20
	Const. on own land	35	13

FIGURE 11.3   *(Continued)*

(ii) Removing excessive checks and balances and keeping it to minimum

(iii) Dispensing with permission for immovable property transaction

But the organization was not ready to adopt these changes at that point of time, as it needed compliance of statutory requirements and induction of enabling (IT) technology. They thought it prudent to implement these in the future.

The suggested process simplification is not the "utopian state" or "best practice approach," but shows considerable scope for improvement.

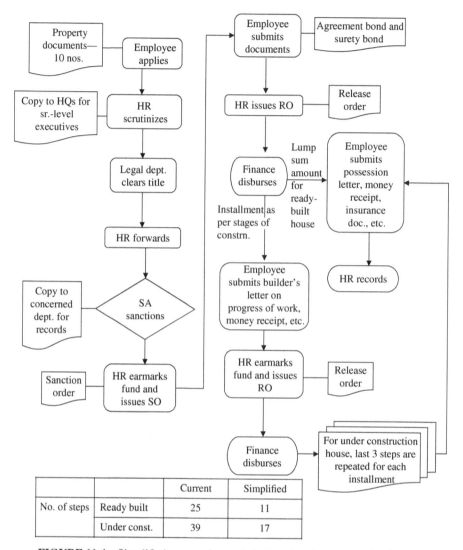

FIGURE 11.4   Simplified process for ready-built and under-construction house.

11.5.2   Official Tour/Duty Travel

The employees going on official tour, also called as "duty travel," seems to be an innocuous event. But the business process for tour approval, drawing advance, ticketing, hotel booking, and related activities are found to be cumbersome and time consuming. There is an opportunity for process simplification and improvement.

11.5.2.1 Existing Process: The current business process for employees going on outstation official tour/duty travel is a lengthy process, which is described in the following paragraphs:

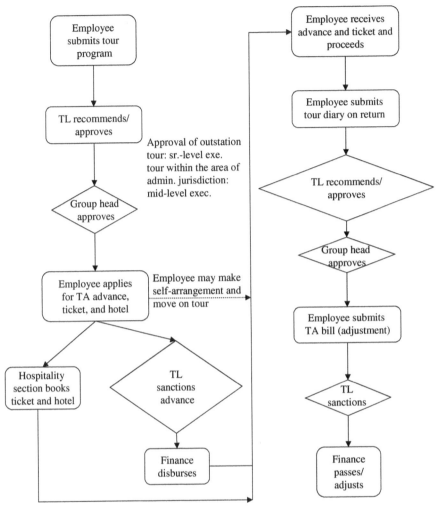

TL: team lead;   TA: travel allowance

FIGURE 11.5   Duty travel: current process.

(a) The employee has to fill out four forms and obtain approval from the group head before proceeding on outstation official duty. These pretravel forms are as follows:
   (i) Tentative travel program
   (ii) Travel advance
   (iii) Requisition for tickets
   (iv) Requisition for hotel booking

After returning from the outstation official tour, the employee has to submit two more forms and obtain approval of the group head. The two posttravel forms are as follows:
   (i) Travel diary (actual)
   (ii) Travel allowance adjustment

Needless to say, the existing process is tedious and taxing, as it involves **14 steps** as shown in the workflow diagram in Figure 11.5.

(b) Even drawing of travel expense advance is not as simple as it seems. Employee has to furnish surety bond, if duration of his/her employment in the company is less than the specified period (e.g., 3 years).

(c) Filling out of these six forms is time consuming, especially the "official tour/ duty travel adjustment" form, which requires minute details of itinerary such as date and time of departure and arrival at each port, mode of transportation (air, rail, and road), type of local conveyances used, and complex categorization of daily allowances based on several criteria, such as employee level, cost of living of the visiting port, types of boarding (self-arranged, hotel, and company guest house), and so on. The employee has to furnish documentary proof as well as self-certification for the aforementioned information. It takes a minimum of 30 min or more to fill out the duty travel adjustment form and is a cumbersome task.

11.5.2.2 Simplified Process: The simplified process is shown in Figure 11.6:

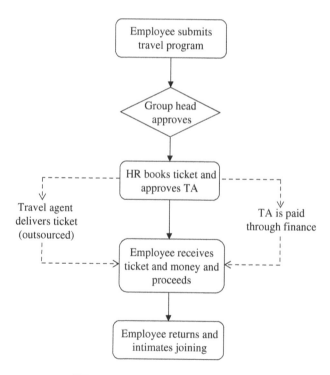

	Current	Simplified
No. of steps	14	5

FIGURE 11.6   Duty travel: simplified process.

(a) *The simplified process contains 5 steps against the existing 14 steps, and employee has to fill out only one form instead of six.* The information required from the employee in the new form are minimal and limited to employee name and identity number, place of visit, date/duration, and purpose of visit, which would not take more than 2 min to fill out.

(b) The employee submits the duty travel form, group head approves, and HR books ticket (through outsourced travel agency) and pays travel expenditure to employees through finance based on per diem rate.

(c) After returning from the outstation tour, the employee intimates "date of joining duty" and "days spent on duty." Accordingly, adjustments are made by finance/HR. No additional adjustment bill is required to be submitted by the employee.

11.5.3 No Objection Certificate for Passport

As per the requirement of passport authority/government, employees of state-owned companies in some countries are required to obtain "no objection certificate (NOC)" from their employer for acquiring passport. Passport is an authentic document of an individual's identity and acquiring it is the basic right of a citizen. Therefore, the process of issuing NOC for passport needs drastic simplification.

11.5.3.1 Existing Process: The existing process for obtaining NOC in many state-owned companies is a tedious process, which is described as follows:

(a) The employee has to furnish the following information and documents along with request for NOC:
   (i) Purpose of acquiring passport, which necessarily be overseas visit
   (ii) Means and proof of financing proposed trip
   (iii) Furnishing surety for outstanding loans
   (iv) Clearances from different departments in the organization
       The current workflow process for issuing NOC in the company is shown in Figure 11.7.

(b) The current process contains **14 steps**, which involves excessive paper work, documentation, and clearances from various departments in the organization such as finance, HR, legal, vigilance, D&A (disciplinary and antecedent), and so on.

11.5.3.2 Simplified Process: The simplified process is shown in Figure 11.8, which contains only **two steps**:

(a) NOC may be issued on request from employees after due verifications by HR. For this purpose, the details of employees including comments/clearances from "disciplinary" and other departments may be made available in the computer system.

(b) The list of "reported against" (or red flag) may be introduced and updated regularly, so that it can be instantly verified by the concerned HR officials.

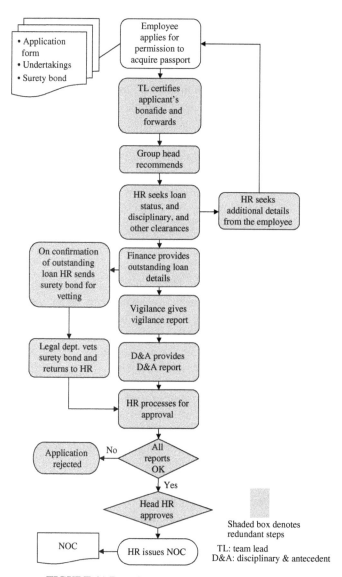

FIGURE 11.7 NOC for passport: current process.

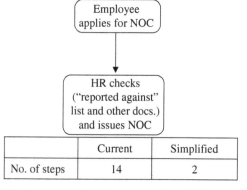

	Current	Simplified
No. of steps	14	2

FIGURE 11.8 NOC for passport: simplified process.

11.5.4 Permission for Visiting Abroad

Many archaic government rules and procedures that are more than 100 years old are still in vogue in many state-owned companies. Many such rules and processes are not conducive to modern-day business and need simplification or dispensation. For example, the employees of state-owned companies need to obtain permission from the organization for visiting abroad, even if it is a personal tour.

11.5.4.1 Existing Process: The existing process of seeking permission for visiting abroad is tedious (and is somewhat similar to obtain NOC for passport, as explained in the previous subsection 11.5.3):

(a) Information on a number of issues are asked for and examined as follows:

(i) Employee's current assignment and its sensitiveness

(ii) Likelihood of employee resigning

(iii) Training undergone by employee and safeguarding company's interest

(iv) Furnishing of surety bond for outstanding loan

(v) Undertaking by employee for not extending leave and returning to the country after visit

(vi) Clearances from disciplinary and other departments

(b) The existing process for obtaining permission for visiting abroad contains 15 steps and involves excessive paper work, documentation, and clearances from different departments that serve the purpose of policing rather than value adding. Besides, the employee has to inform the place, purpose, date and duration of visit, likely expenditure to be incurred, means and proof of fund, and so on. The workflow diagram is similar to that of obtaining NOC for passport (Fig. 11.7).

Most of these information and steps in the current process are not pertinent to majority of employees and have little relevance to the present-day business environment.

11.5.4.2 Simplified Process: The practice of seeking permission for visiting abroad may be done away with. It is unwarranted for personal trip and redundant for official/business trip that, in any case, is approved by the group head. In case of statutory requirement, intimation should be sufficient. This would free the existing resources for better utilization and ease the procedural inconvenience faced by the employees.

11.6 SUGGESTIONS FOR IMPROVEMENT

Some real-life examples of process simplification have been shown in this work. Similar exercise may be carried out for other business areas in an organization. The following suggestions would help in enhancing organizational efficiency:

(a) The simplified processes suggested for availing housing loan (Section 11.5.1.2) and duty travel/official tour (Section 11.5.2.2), obtaining NOC for passport (Section 11.5.3.2), and permission for visiting abroad (Section 11.5.4.2) may be followed to improve organizational efficiency.

(b) Many E&P companies have enterprise resource planning (ERP) system nowadays for business transactions and dealings. Such system can be used to adopt or align with the best practice processes. ERP system provides opportunity and stimuli to simplify processes by

- Centralized processing and shared services
- Eliminating multiple layers and paper trails
- Removing redundant and nonvalue-adding steps
- Mass storage and quick retrieval of data/information

(c) There are several other business processes for reimbursement of small amount claims, such as briefcase, spectacles, swimming, health club, telephone, Internet, professional membership, and so on, which are also cumbersome. These small amount claims may be rationalized keeping in mind process simplification, employee/customer satisfaction, and system optimization.

(d) Rationalize checks and balances, especially for petty claims, loans and advances, and so on.

(e) Employee claims may be processed online and eliminate paper trail; claims requiring original receipt may be supplemented accordingly.

(f) Outsource noncore services/activities such as housekeeping, canteen, travel booking, maintenance of office/colony/guesthouse, logistics support, and so on.

(g) Benchmark various process/service times and use them as performance measure.

(h) Sometimes, it may not be possible to adopt the best practices or attain utopian state of process simplification for the sake of compliance of statutory requirements and government rules. Nevertheless, such processes can be simplified satisfying statutory needs and aligning with the best practice approach to some extent.

11.7 SCOPE FOR FURTHER WORK

Some real-life business process simplification exercises have been demonstrated in this chapter, which would help the practitioners to learn the nuances of process simplification. They can develop simplified processes following the steps described in this exercise. Similar work may be carried out in other functional areas such as the following:

- **Supply chain functions**: Procurement and contracting process, annual planning process, tendering process; various materials management processes including inventory, stock replenishment, disposal, maintenance, verification, and so on; and prequalification, vendor relation, and so forth.
- **Finance and accounts functions**: Accounts payable and receivable, cost accounting, payroll accounting, expense reporting, budgeting process, capital expenditure and administration, forecast, internal audit, and so on.
- **HR functions**: Recruitment and induction process, training and development process, performance review, manpower planning, compensation and benefits, personal claims and settlement process, and so forth.

These are some indicative lists of areas where process simplification studies may be carried out to improve organizational efficiency. However, such studies may be undertaken as per the need of an organization, which may be identified and prioritized based on assessment.

11.8 CONCLUSION

Needless to say, the business process simplification is an inexpensive but powerful aid to identify and remove inefficiencies in the system. Business process simplification may be undertaken in various kinds of system, such as supply chain management, finance, HR, planning, and others. The old systems and processes, which are not conducive to the present-day business, are still continuing in many enterprises causing impediments to efficient functioning of the organization. The need for process simplification should arise from within the organization; accordingly, opportunities for improvement in various business processes may be identified, prioritized, and initiated.

In this study, some HR processes, namely, drawing housing loan, official tour/duty travel, obtaining "no objection certificate" for passport, and permission for visiting abroad, have been illustrated to show the extent and varieties of process simplification and understanding intricacies involved.

The simplified processes have resulted in notable improvement in quality of service from sloppy to competitive, reduced processing time, and improved employee satisfaction. It eliminated redundant and nonvalue-added steps, freed up scarce resources (manpower), and improved awareness about the process and system. The suggestions for improvement are provided in Section 11.6.

The advantage of enabling technology/ERP system, which is in place in most of the large organizations, may be harnessed for process simplification. This would help in adopting and aligning with the best practices. The business process simplification exercises may be undertaken in-house, which would not require additional resources or major changes in policy but would offer tremendous benefits improving organizational efficiency.

Chapter 12 develops a quantitative model to optimize base oil price for a country taking into account the sensitivity and complexities involved in oil pricing, which is dependent not only on economic criteria but also on a number of social, political, and geopolitical issues that strongly influence it. This is a demonstrative model that can be customized by incorporating additional variables and studying the effect of various scenarios under dynamic circumstances.

REVIEW EXERCISES

11.1 How would you identify opportunities for business process simplification? Give some examples where business process simplification would bring improvement in your organization.

11.2 What are the steps you would follow for business process simplification?

11.3 What is lean methodology? How would you apply it to process simplification?

11.4 What are the differences between business process simplification and business process redesign/reengineering? Give some examples.

11.5 What are the benefits of business process simplification?

11.6 How does business process simplification improve organizational efficiency?

REFERENCES

[1] Hammer, M., Reengineering Work: Don't Automate, Obliterate, Harvard Business Review, Volume 68, Number 4, July–August, 1990, pp.104–113.

[2] Hammer, M. and Champy, J., Reengineering the Corporation: A Manifesto for Business Revolution, Harper Business, New York, 1993.

[3] Hammer, M. and Stanton, S.A., The Reengineering Revolution: A Handbook, Harper Business, New York, 1995.

[4] University at Albany, SUNY, Process Management and Process Oriented Improvement Programs, Business Process Modeling, Simulation and Design, Fall 2006, available at: www.albany.edu/acc/courses/acc630.fall2006/ch02.ppt (accessed on March 23, 2016).

[5] Davenport, T.H., Process Innovation: Reengineering Work through Information Technology, Harvard Business School Press, Boston, MA, 2013.

[6] Davenport, T.H. and Short, J.E., The New Industrial Engineering: Information Technology and Business Process Redesign, Sloan Management Review, Volume 31, Number 4, 1990, pp. 11–27.

FURTHER READING

Carr, D.K. and Johansson, H.J., Best Practices in Reengineering: What Works and What Doesn't in the Reengineering Process, McGraw-Hill, Inc., New York, 1995.

Crosby, P.B., Quality Is Free, New American Library, New York, 1979.

Crosby, P.B., Quality without Tears: The Art of Hassle-Free Management, McGraw-Hill, New York, 1984.

Deming, W.E., Out of Crisis, MIT Center for Advanced Engineering Study, Cambridge, MA, 1986.

Evans, J.R. and Lindsay, W.M., The Management and Control of Quality, 4th edition, South-Western, Cincinnati, OH, 1999.

Garvin, D.A., Competing on the Eight Dimensions of Quality, Harvard Business Review, Volume 65, Number 6, Nov–Dec, 1987, pp. 101–110.

Garvin, D.A., Managing Quality, Free Press, New York, 1988.

Goetsch, D.L. and Davis, S., Implementing Total Quality, Prentice-Hall, Upper Saddle River, NJ, 1995.

Hall, R., Attaining Manufacturing Excellence, Dow-Jones Irwin, Burr Ridge, IL, 1987.

Harrington, H.J., Business Process Improvement: The Breakthrough Strategy for Total Quality, Productivity, and Competitiveness, McGraw-Hill, New York, 1991.

Juran, J.M., Juran on Planning for Quality, Free Press, New York, 1988.

Manganelli, R.L. and Klein, M.M., The Reengineering Handbook: A Step-by-Step Guide to Business Transformation, Amacon, New York, 1994.

Peppard, J. and Rowland P., The Essence of Business Process Reengineering, Prentice-Hall Ed, New York, 1995.

Schonberger, R.J., World Class Manufacturing: The Lessons of Simplicity Applied, Free Press, New York, 1986.

Sharp A. and McDermott P., Workflow Modeling: Tools for Process Improvement and Application Development, Artech House Publishers, Boston, MA, 2001.

Womack, J.P. and Jones, D.T., Lean Thinking: Banish Waste and Create Wealth in Your Corporation, Simon and Schuster, New York, 1996.

Useful Links

Hoffman, M.V. and Malhotra, N., Process Simplification—The Simple Way!, available at: www. lexjansen.com/nesug/nesug12/ma/ma06.pdf (accessed on March 3, 2016).

Malhotra, Y., Business Process Redesign: An Overview, available at: www.brint.com/papers/bpr.htm (accessed on March 3, 2016).

Minder, C., Business Process Reengineering, available at: http://slideplayer.com/slide/7075257/ (accessed on March 23, 2016).

Zygiaris, S., Business Process Re-Engineering – BPR, INNOREGIO project, available at: www. adi.pt/docs/innoregio_bpr-en.pdf (accessed on March 3, 2016).

12

OPTIMIZATION OF BASE OIL PRICE USING LINEAR PROGRAMMING

12.1 INTRODUCTION

Oil is an important commodity in today's world, which influences the daily life of every individual, impacts the economic growth and development of every nation, and influences the world economy to such an extent that no other commodity can match. Oil pricing is a sensitive issue; therefore, it is important that oil price should be fair to both producers and consumers for sustainable growth and judicious consumption. In this chapter, we would discuss the importance, sensitivity, and complexities involved in oil pricing and would develop a linear programming (LP) model to optimize base oil price for a country.

Oil pricing is a complex issue and is not governed by economic criteria alone—a host of socio-political and other factors influence it. This study aims to optimize base oil price taking into account the cost and share of domestic oil production and that of imported oil, effect of reserves life on price, effect of substitute to oil, and so on. It develops an optimization model using LP method, which has the capability to study and measure the impact of varying parameters under different scenarios.

Because of complexities involved in pricing a product and the sensitivity and uniqueness of oil pricing, a brief note on pricing in general and oil pricing in particular has been elaborated in the following paragraphs.

12.1.1 Pricing

The survival of an enterprise or business is greatly dependent on its pricing policy, which also influences the balanced growth of both consumption and supply. The pricing decision is mostly the outcome of satisfying and balancing a number of issues including economic, business environment, business maturity and health, and other specificities. The price is a

Optimization and Business Improvement Studies in Upstream Oil and Gas Industry, First Edition.
Sanjib Chowdhury.
© 2016 John Wiley & Sons, Inc. Published 2016 by John Wiley & Sons, Inc.

determinant of revenue, which a firm usually seeks to maximize. As mentioned earlier, price fixation is a delicate issue that is influenced by numerous internal and external factors. The internal factors are predominantly cost management policy including operational, commercial, and financial aspects. The cost is a major determinant factor and indicator of price—it specifies the boundary or resistance point to the lowering of price beyond which the product is not economical or business is not sustainable. The potential external factors are elasticity of supply and demand, competition, company goodwill, purchasing power of buyers, entry barrier, and government policy.

The factors influencing pricing of different commodities are generally not the same. Even the factors influencing pricing of the same commodity may vary from country to country or market to market. Pricing is not an exact science nor does it follow any particular economic model, even though it is influenced by the economic factors. Pricing is more of a judgment based on sound economic principles and reliable information. There is no infallible formula for determining the right price for a product. Every pricing situation is unique and should be explored based on its distinctiveness.

Pricing assumes significance, especially when a new product is launched or when a product enters a new market competing with an existing product. The factors that are generally considered for determining prices of commodities are maximization of profit, promotion of product, long-range welfare and policies, adoptability and flexibility to meet change, price sensitivity, conflicting interests of manufacturer and middlemen, active entry of nonbusiness groups, market penetration, market skimming, early cash recovery, and so on. Needless to say, there are multitudes of forces, causes, and factors that influence pricing of a product. Accordingly, pricing strategy is categorized as follows [1–3]:

- *Price elasticity of demand*: It is the ratio of percentage change in demand to percentage change in price. The degree of price elasticity impacts the level of sales and revenue. If the demand is inelastic, it would not be profitable to reduce price; therefore, policy of price increase would be appropriate (e.g., increase in price of cigarette, salt, petrol, and peak season air ticket does not affect demand). Conversely, if the demand is elastic, policy of price reduction rather than price increase would be profitable (e.g., increase in price of chocolate bar, fast-moving consumer goods, automobiles, etc., influence the demand).
- *Market skimming*: Generally suitable for products with short life cycle that are likely to face competition in the future and are associated with high price and low volume (e.g., jewelry, digital technology, and PlayStation).
- *Premium pricing*: Sometimes, customer behavior does not follow the law of demand. For example, a dearer product is often perceived as superior quality, and higher prices usually increase the snob appeal of the product. These are generally true for the branded, luxury, and exclusive products. Higher prices that increase consumer readiness to buy may sound uneconomical but may not be unrealistic (e.g., status products, first-class air travel, Cunard Cruise, and luxury hotel suite).
- *Value pricing*: Price is based on consumer perception and response from companies to retain sales due to increased competition or other compelling reasons (e.g., value meal at fast-food restaurants—consumers feel great value for his/her money, and companies are compelled to offer such value product to retain market share).

- *Psychological pricing*: It is closely linked with the value pricing and plays greatly with the consumer perception (e.g., product price USD 9.9 instead of USD 10).
- *Target pricing*: A specified profit level is targeted to set the price of a product. It is mostly used by the public utility companies that make huge investment in electricity, gas distribution, and so on. It is also followed in automobile industry.
- *Marginal cost pricing*: It is the cost of producing incremental (one extra or less) item. It allows flexibility and variable pricing structure. Price of the product covers manufacturing cost but not the overhead cost (e.g., hotels and airlines often resort to marginal pricing to fill the capacity to sustain or improve profitability).
- *Penetration pricing*: It is the price set to penetrate the market, usually for launching products into a new market. It is usually set at a low price to secure market share (e.g., cable or satellite TV operators, home phone, cell phone, and retail store products).
- *Cost plus margin*: Price is based on actual cost plus a markup. Sometimes, it may result in overpricing or underpricing.
- *Contribution pricing*: It is the price that ensures coverage of variable cost and a contribution to the fixed cost. In principle, it is similar to marginal pricing.
- *Absorption/full cost pricing*: Price is set to cover both fixed and variable costs in full cost pricing, while in absorption cost pricing, it is set to absorb part of the fixed cost of manufacturing.
- *Transfer pricing*: It is the price at which transactions are made among the associated enterprises or from one part of the company to another.
- *Tender pricing*: Choosing the best value tender price and carrying out the work accordingly. The contracts are usually awarded on the basis of tender price.
- *Predatory pricing*: It is deliberate underpricing of products to prevent new entrants or coerce rivals to withdraw from the market. It is also called destroyer pricing (e.g., competition between Burger King and McDonald's, and PepsiCo and Coca-Cola).
- *Price discrimination*: Different prices are charged in different markets for the same product or services (e.g., price of suburban train ticket differs at different times of the day for the same journey). It is found when each market is impenetrable and having different price elasticity of demand.

12.1.2 Oil Pricing

Oil pricing is markedly different from product pricing in conventional industry, which operates under the framework of deterministic input and assured output. But E&P industry operates under uncertainty wherein realization of investment is uncertain and may prove futile. Even if the discovery is made, it usually takes a long time to develop the field, assess commercial viability of reserves, and commence production. Because of uncertainty involved in the final outcome of investment, oil pricing is a complex, tricky, and challenging affair.

Oil price is greatly influenced by demand for oil, its supply and reserves position, future government policies on exploration and production, royalty and duties, various taxes and levies, and so on. In addition, other factors such as future technological and economic conditions, timing of production, revision of reserves, and so on, influence the pricing of oil.

Discovery of oil is the prime objective of exploration efforts. But it may so happen that instead of oil, only gas is discovered. Even with the production of oil, associated gas is produced. Therefore, it is a complex issue to calculate the cost of production of oil and that of gas separately. However, for simplification the production of oil and gas may be expressed in terms of "oil and oil equivalent gas" based on the heat content equivalence of oil and gas, which are as follows:

$$1000\,m^3\,(MSCM)\text{ of gas} = 0.90\text{ ton of oil equivalent or}$$

$$1000\,ft^3\,(MSCF)\text{ of gas} = 0.1767\text{ barrel of oil equivalent}$$

Oil price influences the economic development of a country—both developing and developed nations alike. In order to draw future plan of a country or industry or enterprise, it is necessary to project future demand for oil and more importantly oil price. But reliable forecast of oil price is elusive; it is an enigmatic issue and no long-term prediction has proved to be infallible; it provides at best a trend. There are various groups and agencies engaged in this task all over the world, as every nation has a stake in it. A large number of studies on oil price forecast and demand for oil have been made by various groups and agencies. Some of the notable groups and agencies whose predictions are somewhat reliable and are often referred to by many are International Energy Agency (IEA), US Energy Information Administration (EIA), PIRA Energy Group, Purvin & Gertz, Wood Mackenzie, Cambridge Energy Research Associates (CERA), Energy Security Analysis Inc. (ESAI), and others.

12.1.3   Oil and Gas Accounting

Oil and gas accounting is somewhat different from the conventional accounting system with some specificities. Therefore, a short note on oil and gas accounting is presented in this section for the convenience of the readers.

Oil and gas accounting has evolved over the years to meet the unique nature of upstream business, diverse need, and atypical challenges of E&P sector. An E&P company deals with nonrenewable products and assets, which are associated with high risk, huge investment, and long gestation period. Oil and gas accounting relates to four basic costs, namely, acquisition cost (mining lease), exploration cost, development cost, and production cost. There are two methods of oil and gas accounting, namely, Successful Efforts Method and Full Cost Method. The basic differences between these two methods are related to capitalizing or expensing the incurred cost, and the size of the cost center over which costs are accumulated and amortized.

The cost center under *Successful Efforts Method* is a lease area, field, or reservoir and is much smaller than the cost center under Full Cost Method, which is generally a country. Under Successful Efforts Method, a direct relationship is required between costs incurred and resources discovered. Consequently, only successful searching costs that directly result in the discovery of proven reserves are considered as the cost of finding oil or gas and are capitalized. The costs incurred in finding, acquiring, and developing reserves are typically capitalized on a field-by-field basis. The capitalized costs are allocated to commercially viable hydrocarbon reserves and are depleted on a field-by-field basis as production starts. The unsuccessful searching costs that do not result in an asset are expensed with. In summary, in Successful Efforts Method, the cost center is normally not larger than a field, and the acquisition and development costs are fully capitalized, and exploration cost is partly capitalized (drilling exploratory, appraisal, and test wells) and partly expensed (G&G, unproved properties, dry hole/well, etc.).

In contrast, *Full Cost Method* considers both successful and unsuccessful costs as capitalized, even though the unsuccessful costs have no future economic benefit. All costs incurred in searching, acquiring, and developing the reserves in a large geographic cost center or pool are capitalized. A cost center or pool is typically a country; the cost pools are then depleted on a country basis as production starts. If exploration efforts in the country or the geological formation are wholly unsuccessful, the costs are expensed with. In short, in Full Cost Method, the cost center is typically not smaller than a country, and acquisition, development, and exploration costs are fully capitalized [4].

12.2 OBJECTIVES

Oil pricing is a sensitive and complex issue that goes well beyond the economic criteria. The high risk and uncertainty involved in E&P industry make it a difficult task. In addition, geopolitical, social, and political issues associated with it make it even more challenging. It is often observed that oil price fixed by the state or fixed by oil-exporting countries defies consistency. This might also hold true for unregulated price regime. In fact, oil price should be fair to producers/suppliers as well as consumers for balanced economic growth and maintaining a healthy demand–supply relation. Toward this end, the objectives of this study are as follows:

(a) To optimize base oil price for a country taking into account the cost and share of domestic oil production and that of oil import, reserves life, substitute price, and so on.
(b) To maximize profitability and drive oil production through plowing back profit to E&P activities for sustainable growth
(c) To study the effect of various internal and external factors on oil price, such as cost, profitability, reserves life, international oil price, substitute price, domestic oil production, oil import, and so on.
(d) To develop a model framework that would aid decision making with varying parameters under different circumstances and scenarios

The cost of production of oil and gas is very low in most of the oil-exporting countries, but it fetches enormously high price in international market, which apparently defies economic logic. International oil price is largely controlled by oil-exporting countries by regulating supply. Oil revenue is the main source of national income of these countries. Domestic oil price is heavily subsidized in these countries. On the other hand, oil import drains out the scarce foreign exchange reserves and is a huge burden on the economy of oil-importing nations.

This exercise tries to optimize base oil price for a country taking into account the share and cost of domestic oil production and that of imported oil and other factors. The model is applicable to both oil-importing and oil-exporting countries and can be easily controlled by varying few important parameters, such as percentage demand met through domestic oil production and through oil import, profitability, price multiplier from reserves life, and so on.

12.2.1   Methodology

There are various methods for determining the price of a commodity, and textbooks are replete with them. But oil pricing is a sensitive and strategic issue affecting the entire country, and thus necessitates utmost care and caution. Determination of oil price for a country is usually done by economists through employing various techniques like econometric modeling, cost plus, internal rate of return, and so on. But this can also be done through other optimization methods such as LP, which has the ability to solve complex mathematical problems with relative ease.

This exercise uses LP method for optimizing base oil price and studying the effect of varying parameters on price. LP has the ability to modify the problem (i.e., changing objective function, constraints, parameter values, etc.) and obtain alternative solutions rather quickly with no extra cost. Besides, its versatility, user friendliness, and availability of computer software packages for solving large and complex problem make it a worthy choice. LP method is further discussed in Section 12.3.

12.3   A BRIEF NOTE ON LP

LP is a mathematical optimization technique that deals with linear objective function, linear equality/inequality constraints, and multiple decision variables. Its application is widespread in the business world and is extensively used in the industrial, transportation, energy, oil refinery, manufacturing, telecommunication, and service sectors. It is a useful method for solving diverse planning problems like resource allocation, assignment, scheduling, routing, design, and a host of varied and complex issues. It is also used in the field of science, engineering, management, economics, and others.

The LP technique was initially developed by Russian mathematician Leonid Kantorovich in 1937/1938 to allocate economic resources for maximum output, as a task assigned to him. He conceived that it is a mathematical problem of maximizing a linear function under multiple constraints. This technique was also applied to optimize warfare logistics, minimize cost of army operations, and inflict maximum loss to the rival forces during World War II. But it was kept secret until the **simplex algorithm** was published by George B. Dantzig in 1947 [5]. Simplex algorithm is a popular and powerful technique for solving a wide variety of LP problems and is extensively used nowadays. In postwar, there has been a proliferation of the use of LP method in industries. Notable examples of the application of LP method include airline crew scheduling, blending and material balance in oil refinery, shipping and telecommunication networks, production planning and resource allocation in manufacturing industry, financial portfolio selection in stocks and bonds, and so on.

LP technique has been used in this exercise to optimize base oil price, which is important for a country and for E&P companies. Although LP is used extensively in various fields, its application for fixing or formulating prices is limited; notwithstanding, most of the LP problems have objective function that involve either maximization of profit or minimization of cost. An LP model is generally formulated and represented in the following nomenclature:

Objective function $\rightarrow$ maximize or minimize: $z = c_1 x_1 + c_2 x_2 + \cdots + c_n x_n$

Constraints $\rightarrow$ subject to

$$: \quad a_{11} x_1 + a_{12} x_2 + \cdots + a_{1n} x_n \left(\leq, =, \geq\right) b_1$$
$$: \quad a_{21} x_1 + a_{22} x_2 + \cdots + a_{2n} x_n \left(\leq, =, \geq\right) b_2$$
$$: \quad a_{m1} x_1 + a_{m2} x_2 + \cdots + a_{mn} x_n \left(\leq, =, \geq\right) b_m$$

where

x_j = decision variables
b_i = constraint levels
c_j = objective function coefficients
a_{ij} = constraint coefficients

LP models are based on the assumptions of continuity, certainty, linearity, additivity, divisibility, and so on. More specifically, these assumptions are as follows:

   (i) Parameter values are known with certainty.
   (ii) Variables are continuous and can take any value within the feasible region.
  (iii) There are no interactions between decision variables.
  (iv) Objective function and constraints are linear.
   (v) It deals with single objective.

An LP model is usually solved by the simplex method, which tests the adjacent vertices on the boundary of the feasible set in a sequential manner. The process is reiterated with incremental change in the objective function until an optimal solution is found or it becomes infeasible. LP problems can also be solved by graphical method, but it becomes a bit lengthy, tedious, and difficult, if decision variables are more than two. Simplex, on the other hand, is an efficient, elegant, and powerful method with a wide range of applications in various fields. Following the simplex algorithm, many other algorithms have been developed in subsequent years, which differ in approaches for solving LP problems. These are criss-cross algorithm, Khachiyan's ellipsoid algorithm, Karmarkar's interior-point projective algorithm, approximation algorithm, and so on.

As we proceed further, we would get familiarized with the other important features of LP such as model formulation, sensitivity analysis, duality, shadow or dual price, reduced cost or opportunity cost, slack or surplus variables, and others.

There is a plethora of books available on LP and mathematical optimization. The interested readers may refer to Refs. [6–21] at the end of the chapter to learn more about LP and its applications.

The formulation of LP model for this exercise is discussed in Section 12.4.

12.4 FORMULATION OF THE MODEL

The model has been formulated taking into account the cost and share of domestic oil production and that of imported oil, profitability, reserves life and its effect on oil price, substitute to oil and its dynamics, international oil price and its effect on domestic production, and so on. More on this have been discussed in Section 12.4.8. The formulation of the model is discussed in Sections 12.4.1–12.4.8.

12.4.1 Revenue

Revenue is the earning of an enterprise and is the source of various funds and activities, which influences the health of the organization to a great extent. It is the product of the quantity of oil and gas produced/sold and the price of oil and gas.

It comes mainly from three sources, namely, revenue from oil, revenue from gas, and miscellaneous revenue. Miscellaneous revenue is earned from the sale of value-added products derived from oil and gas such as LPG, NGL, condensate, wax, and so on, and accounts for a tiny fraction of the total revenue for the instant case:

$$Q_1 x_{oil} + Q_2 x_{gas} \geq b_1$$

where

Q_1 = quantity of oil produced/sold
Q_2 = quantity of gas produced/sold
x_{oil} = oil price
x_{gas} = gas price
b_1 = revenue

The quantity of gas produced/sold may be expressed as oil equivalent gas:
$1000\,m^3$ (MSCM) of gas = 0.90 ton of oil equivalent,

$$Q = Q_1 + 0.0009 Q_2 \text{ or}$$

$1000\,ft^3$ (MSCF) of gas = 0.1767 barrel of oil equivalent,

$$Q = Q_1 + 0.0001767 Q_2$$

Therefore,

$$Q x_1 \geq b_1; \quad \text{Or}, \ x_1 \geq x_1 \geq \frac{b_1}{Q} \tag{12.1}$$

where

Q = quantity of oil and oil equivalent gas
x_1 = unit oil price

12.4.2 Cost

Cost is a major determinant factor as well as an indicator of price; it represents a resistance point to the lowering of price. The effectiveness of an organization is reflected to some extent on its ability to hold down the overall controllable cost. It is desirable to minimize the cost of an organization through better utilization of resources and prudent economic and operational decisions. E&P business is associated with high-cost activities, and holding down the base cost is one of the main concerns of E&P companies across the world. *The total cost is the sum of statutory cost (government taxes, duties, and levies) and operating cost that includes the cost for amortization, depletion, depreciation, impairment loss, and operational cost, besides general administrative expenses*:

$$\text{Total cost} = \text{statutory cost} + \text{operating cost} \begin{pmatrix} \text{including amortization, depletion, depreciation, impairment loss,} \\ \text{G \& A, and other operational costs} \end{pmatrix}$$

Statutory cost consists of various taxes, duties, and levies imposed by the government such as royalty, levies, and local and central taxes. In addition, corporate tax is charged by the government on the profit of the company. An E&P company has little control on the statutory cost imposed by the government, which is a major source of government revenue. The amount of statutory cost varies from country to country and is quite high in oil-exporting countries. It is the main source of income of these countries and accounts for more than 80% of the gross revenue (or over 95% of the total cost).

The analysis of 10 years of cost data of an E&P company in an oil-importing country reveals that the statutory cost accounts for approximately 20% of gross revenue (or ~30% of the total cost). This may be indicative of many other oil-importing nations.

Operating cost *is the sum of amortization, depletion, impairment loss, depreciation, and other operational costs.* **Amortization** means recovery of expenses in part through the year's revenue for those fields, which have either been abandoned or are at the exploratory stage or whose commercial viability is yet to be determined.

Depletion can be defined as the prorated value assigned to the extinction of natural resources, that is, oil and gas. It is difficult to prorate expenditure incurred on exploration, development of field, reserves accretion, and so on, on unit basis for the purpose of long-term plan. It is depleted according to policies followed by the respective E&P companies, which may change from time to time. In addition, there is **impairment loss** that occurs when the cost of holding the asset exceeds its fair market value. In other words, it is the amount by which the carrying amount of an asset exceeds its recoverable amount.

Depreciation is a source of fund and is created by charging wear and tear and obsolescence cost at a predetermined rate.

Other operational costs include the expenditure incurred toward the operation of workover rig, water injection, research and development, foreign contracts, transportation of oil and gas, and so on.

Therefore,

$$Qx_2 \leq b_2; \quad \text{or,} \quad x_2 \leq \frac{b_2}{Q} \tag{12.2}$$

where

b_2 = total cost
x_2 = unit cost of production of oil and oil equivalent gas

12.4.3 Profitability

Profitability is the ratio of net income to revenue and is often considered as the bottom line of a business or commercial venture. It is important to achieve the desired profitability for survival and sustained growth of an enterprise. It is the basic instinct of business and is required to be maximized. Further, profit is the source of fund for future investment and carrying out future E&P activities. Profitability takes into account the total income and expenses of the enterprise or business. The profitability of E&P companies varies widely depending on various factors, namely, reserves life, production capacity, statutory cost, and operating cost.

It was observed from 10 years of data of an E&P company that the net profitability margin (after expensing statutory cost, corporate tax, etc.) is approximately 30%. Considering the uncertainties involved and long gestation period for realization of investment in the E&P sector, the average profitability of 30% is generally accepted:

$$\frac{Qx_1 - Qx_2}{Qx_1} \geq P_f \qquad\qquad \text{where} \quad P_f = \text{profitability ratio}$$

$$\text{or,} \quad 1 - \frac{x_2}{x_1} \geq P_f$$

$$\text{or,} \quad \frac{x_2}{x_1} \leq 1 - P_f \tag{12.3}$$

$$\text{or,} \quad \frac{x_2}{x_1} \leq 0.70$$

$$\text{or,} \quad x_1 \geq \left(\frac{1}{0.7}\right) x_2 \qquad \left(\text{considering } P_f = 0.30\right)$$

$$\text{or,} \quad x_1 - 1.4285 \; x_2 \geq 0$$

12.4.4 Substitute Price

The substitute to oil would gradually penetrate the energy market due to scarcity of oil and rising prices. The substitute to oil may be in the form of alternative sources of energy, such as solar energy, wind power, nuclear energy, hybrid car, biofuel, geothermal energy, coal bed methane, and others.

As long as oil price is less than substitute price (x_3), the share of oil is dominant in the energy market. It would decline with the rise in oil price. It is difficult to predict the break-even price at which substitute would become competitive to oil. This is a contentious issue and there is little consensus among experts on this. In this study, we are primarily concerned with formulating the model to use it as a tool for determining/optimizing base oil price. For this purpose, it is assumed that the share of oil will continue to be dominant in the foreseeable future. Therefore, the substitute price has been set at a high price, say, USD 200 per barrel:

$$0 \leq x_1 \leq x_3$$
$$x_1 - x_3 \leq 0 \tag{12.4}$$

where

$x_3 = 200$ USD per barrel

12.4.5 International Oil Price

International oil price (x_4) is the price at which a country purchases oil from the international market and through bilateral or long-term agreement with oil-exporting countries. The domestic oil production will continue as long as (domestic) production cost is less than or equals international oil price. If international oil price becomes cheaper than the domestic oil production cost, then there will be little incentive to produce oil at home. In such cases, oil import will be dominant, and the share of domestic production will decline. This holds true for oil-exporting countries also. However, purchase of oil from the international market will be governed by the capacity to buy, availability of foreign exchange reserves of the country, and so on.

$$x_1 \leq x_4 \left(\text{where } x_4 \text{ is the international oil price} \right)$$
$$x_1 - x_4 \leq 0 \tag{12.5}$$

12.4.6 Reserves–Production Ratio

Reserves–production ratio (RPR) is the measure of reserves life and is the ratio of recoverable oil reserves to current oil production rate. It is an important ratio for formulating future strategies and indicates how long the reserves would last at the current production rate. A country with high RPR is in a comfortable position in respect of energy availability. On the other hand, a low RPR signifies its vulnerability to outside shock and limited options for maneuverability.

The economic prosperity of a country is closely related to its energy consumption level. The higher the per capita energy consumption, the higher is the economic prosperity of the country. It is desirable to increase oil production capacity for all-round growth of the country. But increased oil production would result in early depletion of reserves, and if produced at a faster rate beyond the optimum level, it would affect the health of the reservoir, which would be detrimental in the long run. In order to strike a balance between the economic growth of a country and the reserves life, the RPR shall be kept at an optimum level. The desired value of RPR varies from country to country depending on the quantum or availability of reserves, production capacity, and consumption rate. For the instant case/country, the desired RPR is 20 years, which is indicative of many other countries.

12.4.7   Price Multiplier from RPR (P_m)

P_m represents the effect of relative abundance or scarcity of oil reserves on the price. In essence, price multiplier (P_m) defines the production supply curve in the unregulated market. It represents the reaction of producers in setting prices according to changing conditions of supply and demand. The nature of price multiplier curve is shown in Figure 12.1, which has been derived from modifying the assumption and work done by Roger F. Naill [22].

Oil price increases with the decrease in RPR. This is because depleted reserves are likely to be replenished either by new discovery or by tapping more costly known reserves. This in effect would push the price higher with dwindling RPR. Conversely, with abundance of reserves and high RPR, cost will come down in the unregulated market. Roger F. Naill has assumed that price never drops below 0.9 times the cost plus margin and has plotted the graph up to RPR/DRPR = 2. But in reality, the supply is regulated by oil-exporting countries with very high RPR (80–100 years); in other words, countries with RPR/DRPR > 3 or 4 control the international oil price. Thus, even with abundance of reserves and high RPR, it is unlikely that the price multiplier would fall below the cost plus margin. This assumption has been made based on the real-world situation for the last few decades, and accordingly RPR/DRPR has been plotted up to 4 in this study (see Fig. 12.1).

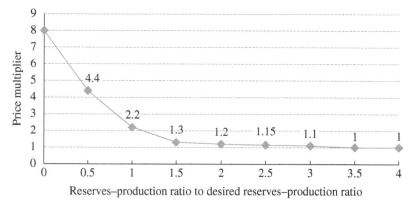

FIGURE 12.1   Price multiplier from reserves–production ratio to desired reserves–production ratio.

12.4.8 Objective Function

The objective of the model is to optimize base oil price for a country. Accordingly, the objective function has two major components, namely, indigenously produced base oil price and imported base oil price. Therefore, the share of domestic oil production and that of oil import are important parameters for apportioning and optimizing base oil price.

12.4.8.1 Indigenously Produced Base Oil Price: *The indigenously produced base oil price is the cost of domestic oil production plus the cost plus margin and factoring in price multiplier from reserves life.*

The maximization of profit is the basic instinct of business and is the source of fund for survival and growth. It is desirable to maximize profit that is dependent on domestic oil production and price. Thus, price maximization would drive profitability and oil production by reinvesting profit into E&P activities. The domestic unit base oil price (x_1) may be derived from the unit base cost (x_2) multiplied by price multipliers (P_m) from RPR divided by one minus net profitability ratio or profit margin:

$$\text{That is, } x_1 = \frac{P_m}{(1 - P_f)} x_2 \tag{12.6}$$

12.4.8.2 Imported Base Oil Price: *International oil price is the prerefinery crude oil price that is traded in the spot market.* Crude oil price in international market depends on types and API grades. In order to avail better profit margin, different types of crude are usually blended into one or two grade(s) by most of the countries.

On the other hand, *imported base oil price may be defined as international oil price (spot price) plus transportation cost, freight, insurance, handling cost, and so on, up to the refinery.* The mode of transportation is usually shipping from sellers' terminal to buyers' port by seaborne oil tankers and through pipeline from the port to refinery.

Therefore,

$$\text{Transportation cost} = \text{shipping cost of freight (from sellers' terminal to buyers' port)} + \text{pipeline tariffs (from port to refinery)}$$

where shipping cost of freight (λ) is expressed in USD/barrel and pipeline tariff (μ) is also expressed in USD/barrel.

A part (γ percent) of the FOB value and freight has been considered to cover insurance, handling, and miscellaneous charges and is expressed in percentage of international oil price (spot price) and quantity of shipping.

Therefore,

$$\text{Imported base oil price} = (1 + \gamma) x_4 + \lambda + \mu \tag{12.7}$$

where

x_4 = international oil price

Most of the oil-importing nations meet part of their oil and gas demand through domestic production, and the rest are imported. The percentage of domestic oil production and import varies from country to country. Many countries are heavily dependent on oil import to meet their demand; some are moderately dependent as they produce part of their demand indigenously, while few countries solely export to meet the world oil demand.

Let β be the share (percentage) of demand met through domestic oil production; thus, $(1-\beta)$ is the share (percentage) of demand met through oil import.

12.4.8.3 Formulated Model: The formulated LP model is summarized as follows:

Objective function $\rightarrow$ maximize $Z = \beta * \left(\dfrac{P_m}{(1-P_f)} * x_2 \right) + (1-\beta) * \left\{ (1+\gamma) x_4 + \lambda + \mu \right\}$

Subject to $x_1 \geq (b_1 / Q)$ or $Qx_1 \geq b_1$

$$x_2 \geq \frac{b_2}{Q} \text{ or } Qx_2 \leq b_2$$

$$x_1 - 1.4285 \quad x_2 \geq 0$$

$$x_1 - x_3 \leq 0$$

$$x_1 - x_4 \leq 0$$

$$x_3 = 200$$

$$x_4 = 107.31$$

$$\text{and } x_1, x_2, x_3, x_4 \geq 0$$

where

x_1 = unit price of oil and oil equivalent gas
x_2 = cost per unit of oil and oil equivalent gas
x_4 = international oil price
x_3 = substitute price
b_1 = revenue
b_2 = total cost
P_f = profitability ratio
P_m = price multiplier from reserves life
β = % demand met through domestic oil production
$(1-\beta)$ = % demand met through oil import
γ = insurance and other costs like handling and miscellaneous charges (%)
λ = transportation (shipping of freight) cost (USD/barrel)
μ = transportation (pipeline tariff) cost (USD/barrel)

12.5   MODEL CALIBRATION AND EXECUTION

The model was calibrated and tested with the data collected from various sources like published annual reports of major E&P companies and published reports from various international bodies, agencies, and so on.

Run 1:   The relevant data of an E&P major in an oil importing country that produces 90% of country's domestic oil production were collected. Domestic production accounts for only 30% of country's oil demand while the remaining 70% are imported. By virtue of its preeminent position, production capacity, size, and potential, the company influences the government decision on domestic oil price.
   The model was calibrated with the actual data of Year N for Run 1 as follows:

Revenue (b_1) = USD 16.9635 billion
Total cost (b_2) = USD 9.9693 billion
Quantity of oil and oil equivalent gas produced/sold (Q_{oeg}) = 829 MBOPD
Profitability (P_f) = 0.3
Price multiplier from reserves life (P_m) = 1
International oil price (x_4) = USD 107.31 per barrel
Percentage demand met through domestic oil production (β) = 30%
Percentage demand met through import of oil $(1 - \beta)$ = 70%
Insurance and other costs like handling and miscellaneous charges (γ) = 0.1%
Transportation (shipping of freight) cost (λ) = USD 2 per barrel
Transportation (pipeline tariff) cost (μ) = USD 1 per barrel

Run 2:   In order to demonstrate the contrast and wide difference in base oil price, cost of production, and profitability, the model was also tested with the data of an *E&P major in an oil-exporting country* that produces over 97% of the country's oil production and influences the decision of determining the domestic oil price.
   Similarly, the model was calibrated with the actual data of Year N for Run 2 as follows:

Revenue (b_1) = USD 19.368 billion (net of royalty and levy (statutory cost) that
   accounts for more than 81% of gross revenue)
Total cost (b_2) = USD 4.199 billion
Quantity of oil produced/sold (Q_{oil}) = 3.181 MMBOPD
Profitability (P_f) = 0.783
Price multiplier from reserves life (P_m) = 1
International oil price (x_4) = USD 107.31 per barrel
Percentage demand met through domestic oil production (β) = 100%
Percentage demand met through import of oil $(1 - \beta)$ = 0%
γ, λ, μ are not applicable, as oil is not imported but exported.

Execution: The formulated model was executed with the help of computer using LP software packages. There are several LP software packages available in the market, and one may refer to any Operations Research or LP textbook (see References and Useful Links) or surf the Internet to find an appropriate one that would justify the cost, time, and effort. The model was executed with two different software packages for both Run 1 and Run 2 to check the uniformity of solutions and unanimity of results. The model was also tested with varying data (for different years: Year $N-1$, Year $N-2$, Year $N-3$, etc.) and changing parameter values. The results are discussed in Section 12.6.

12.6 RESULTS

The results of Run 1 and Run 2 are discussed in detail in the following paragraphs.

Run 1

(a) The output of the model (using one of the popular LP software packages) for Year N is shown in Appendix 12.A.

It may be noted that x_1 and x_2 are decision variables: x_1 represents indigenously produced domestic oil price, and x_2 is the unit cost of indigenously produced oil, and the maximized objective function represents the optimal base oil price. *The results show that objective value or optimum base oil price is USD 91.41 per barrel, unit price for indigenously produced oil (x_1) is USD 56.07 per barrel, and unit cost for indigenously produced oil (x_2) is USD 32.95 per barrel.*

(b) The result provides important information on shadow or dual price, reduced or opportunity cost, and slack or surplus variables. The physical significance of these values/results (Appendix 12.A) is explained in the following paragraphs:

Shadow or dual price denotes sensitivity of profit to resource quantities at optimal level. It is the value per additional unit of a resource at optimal level. It is the ratio of incremental change in optimal objective function to unit increase in the right-hand-side (RHS) value of constraint:

$$\text{That is, shadow or dual price} = \frac{\delta \left(\text{change in optimal objective function}\right)}{\delta \left(\text{unit change in RHS value of constraint}\right)}$$

It represents the maximum price at which one should consider buying additional units of resource. It also represents the minimum price at which one should consider selling units of resource [23].

In reference to Appendix 12.A, for every unit increase in the RHS of constraint (2), the optimal functional value will increase by 0.42855 as it is a maximization problem. Similarly, there is corresponding increase in optimal function value by 0.7007 with increase in RHS of constraint (7).

(c) **Reduced or opportunity cost** is associated with the coefficients of decision variables in the objective function. The value of some decision variables at optimal solution may be zero; these are nonoptimum decision variables. *The value of reduced cost indicates worsening of optimal objective function per unit of non-optimal variable introduced into design.* Each variable in an LP model has an associated reduced cost, which can be positive, negative, or zero. In LP, a variable can only be in the optimal solution, other than upper or lower bound, if its reduced cost is zero. Therefore, if the variable is in the model solution and its reduced cost is zero, it would not be affected by upper or lower bound.

In reference to Appendix 12.A, the reduced cost of variables is zero, which means the variables are in the optimal solution.

(d) The **slack or surplus variables** represent the scarcity or excess availability of the resources. It tells about the unutilized resources or the requirement of the additional resources and gives an insight to its possible transfer or mobilization.

LP problems are generally solved by the simplex method, which requires compliance of equality condition in constraint equations. The inequalities in constraint equations are changed to equalities by adding or subtracting the required amount of resources, which is denoted by a variable called "slack or surplus" variable.

A variable added to a less-than-or-equal-to constraint is called "slack" variable. It represents unused or idle resources. Similarly, a variable added to a greater-than-or-equal-to constraint is called "surplus" variable. It represents the extra amount of resources being utilized over its minimum requirement.

In reference to Appendix 12.A, the constraints (1) and (2) have surplus and slack variables, respectively, with value zero. This shows that these constraints are exactly satisfied. The constraints (4) and (5) have slack variables with values $(r4, r5) = (143.93, 51.24)$. The surplus variable value obtained is $r3 = 9.00$.

(e) The model was also tested with the help of another LP software package (using *dual* simplex for phase 1 and *primal* simplex for phase 2) to test the consistency of the results. There was unanimity in results for both cases, which are shown in Appendices 12.1.1 and 12.1.2.

(f) Similarly, the model may be run for Year $N-1$, Year $N-2$, Year $N-3$, and Year $N-4$, and the model-computed base oil price may be compared with the actual base oil price (regulated or unregulated as the case may be) set by the government. This would help in checking the rationality of fixing base oil price in the past.

Considering the number of exogenous variables that influence oil price and the compulsions of the state due to sociopolitical reasons, variations are expected. However, the extent of variations differs from country to country.

(g) It was observed that the model-computed base oil price (i.e., maximized objective function) was increasing over the years, which is in sync with the trend of international oil price. It is generally found that oil price increases with the rise in demand and depletion of reserves, which conforms to the nature of pricing of the finite natural resources. This may be attributed to the fact that with more and more discovery and depletion of easy reserves, the search for oil is directed to unfavorable, hostile, and difficult terrain and at greater depth, which only add to the cost of exploration and production of oil and gas. Further, at the initial stage oil comes out (of its own) with natural drive, as formation pressure is high. But with the passage of time and maturity of the field, the pressure drops and necessitates deployment of artificial lift and secondary and tertiary methods of recovery. All these increase the cost of production. The development of technology is aimed at bringing down the cost of exploration and production, but technology is unable to match the rising cost of depletion and insatiable global demand for oil.

Run 2: In order to demonstrate the wide variability in base oil price, cost of production, and profitability in oil-exporting and that of oil-importing countries, the model was run with the data of a major E&P company in an oil-exporting country. The model calibration has been discussed in Section 12.5. The results of Run 2 are discussed in the following paragraphs:

(h) The result of Run 2 is furnished in Appendix 12.B, which clearly shows that the objective value or base oil price is USD 16.43 per barrel, which is nearly one-sixth of an oil-importing country's price; unit cost of production is USD 3.616 per barrel, which is about one-tenth of an oil-importing nation's cost of production; and profitability is many times higher compared to oil-importing nations.

(i) The result also depicts shadow or dual price, reduced or opportunity cost, and slack or surplus variable values. The physical significance of these has been explained in list items "b–d" and also in Sections 12.6.1 and 12.6.2.

  In reference to Appendix 12.B, the results are further explained as follows:

  Shadow or dual price: For every unit increase in the RHS of constraint (2), the optimal functional value will increase by 4.5454 as it is a maximization problem.

  Reduced or opportunity cost: The reduced cost of variables is zero, which means the variables are in the optimal solution.

  Slack or surplus variables: The constraints (1) and (2) have surplus and slack variables, respectively, with value zero. This shows that these constraints are exactly satisfied. The constraints (4) and (5) have slack variables with values $(r4, r5) = (183.32, 90.63)$. The surplus variable value obtained is $r3 = 0.243$.

(j) The model was also tested with another LP software package to check the consistency of results. The results using both these software packages were found consistent and uniform. The results of this run are shown in Appendices 12.B.1 and 12.B.2.

(k) Similar to Run 1, the model may be tested with the data of different years, say, Year $N-1$, Year $N-2$, Year $N-3$, and Year $N-4$, and the model-computed base oil price may be compared with the actual base oil price fixed by the government.

Many important postoptimal queries can be answered from the numerical information of the final simplex iteration.

12.6.1 Sensitivity Analysis

Sensitivity analysis is essentially the postoptimal analysis that tests the effect (sensitivity) of optimal solution with respect to the following:

(i) Change of coefficients in objective function
(ii) Change of coefficients in constraint equations
(iii) Change of constant terms in constraints (i.e., RHS values)

The business world is dynamic—the environment in which the business operates is ever changing, which may necessitate change of coefficient values in objective function and/or in constraints.

Furthermore, the initial values of coefficients in objective function and in constraints are sometimes best estimates and may not be accurate. Some coefficient values need to be fine-tuned with trial-and-error method. Sensitivity analysis or "what-if" test tries to address all these issues. It determines the effect on optimal solution due to changes in coefficient values in objective function and in constraints.

(a) Sensitivity analysis tells how far the input parameters (coefficients) can be varied without causing abrupt or significant changes in a compound optimal solution and helps to understand the range of validity of the answer. It would also show the range of variation in each objective function coefficient and RHS constant over which the current basis remains feasible and optimal.

The **range of optimality** for each objective function coefficient is the range of values over which the current solution will remain optimal. The optimum solution may be improved upon by expanding the feasible region. This can be done by increasing the upper bound and/or decreasing the lower bound of feasible region, that is, relaxing the constraints. It is better to focus on the objective function coefficients with narrow range of optimality and/or the coefficients that are near the edges of the range.

The objective and range of variability for sensitivity analysis for Run 1 and Run 2 are shown in Appendices 12.3 and 12.4, respectively.

In reference to Appendix 12.C and 12.4, there are lists of all four variables with four types of limits, namely, "from," "till," "from value," and "till value." The coefficient of objective function value may vary in between these limits obtained so that the solution remains the same. The variable X_2 has a "from" value zero to "till" value positive infinity. The coefficient of objective function value of variable X_2 can vary from zero to positive infinity, but the solution will not change. The variable X_2 will become the "till" value, that is, zero, if the coefficient in the objective function of X_2 reaches the "from" value, that is, zero.

(b) The effect of change in various coefficient values can be studied by carrying out sensitivity analysis. For example, the effect of varying β (i.e., share of demand met through domestic oil production) on base oil price (objective value) for Run 1 is shown in Figure 12.2.

It may be seen from Figure 12.2 that with the rise in β, base oil price decreases neutralizing the effect of high international oil price.

Similarly, the sensitivity analysis may be carried out to study the effect of changing various coefficient values and constant terms, such as profitability (P_f), cost (b_2), price multiplier from reserves life (P_m), international oil price (x_4), and others.

Furthermore, the following points may be noted for sensitivity analysis of the RHS value of constraints [24]:

(i) The optimal solution changes with the change in the RHS of a *binding constraint*.
   • The size of feasible region changes with the RHS values of binding constraints.
   • The optimal objective function improves with the increase in size of feasible region.
   • The optimal objective function worsens with the decrease in size of feasible region.

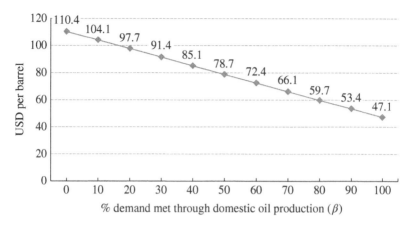

FIGURE 12.2   Base oil price with varying β (sensitivity analysis).

(ii) The optimal solution remains unchanged, if any change in the RHS of a *nonbinding constraint is less than its slack or surplus.*

Sensitivity analysis is a useful and efficient decision-making tool/method for policy makers and managers to test the effect of changes, evaluate various alternatives, and choose the best option. It also provides useful economic and operational information. Some "what-if" tests may be quite complex and tedious to solve, but with the availability of powerful LP software packages nowadays, it is easier and quicker to obtain satisfactory solutions.

12.6.2   Dual Problem

Linear program is essentially formulation of optimization problem. The initial LP model/ formulation is typically called *primal* (like the one developed in this exercise – see Section 12.4.8.3). Every linear optimization problem has an associated linear program called *dual*. By solving one of these problems, the other is solved automatically. If a primal LP problem has an optimal solution, so does its dual (i.e., Value$_{primal}$ = Value$_{dual}$), which is also known as *strong duality* [25].

The formulation of the dual of our original (primal) LP problem is shown in the following paragraph:

Objective function $\rightarrow$ minimize $w = b_1 y_1 + b_2 y_2 + 200 y_6 + 107.31 y_7 + (1 - \beta) * (\lambda + \mu)$

Subject to $-Q y_1 - y_3 + y_4 + y_5 \geq 0$

$$Q y_2 + 1.4285 y_3 \geq \beta * \frac{P_m}{(1 - P_f)}$$

$$-y_4 + y_6 \geq 0$$

$$-y_5 + y_7 \geq (1 - \beta)(1 + \gamma)$$

$$y_1 \geq 0$$

$$y_2 \geq 0$$

$$y_3 \geq 0$$

$$y_4 \geq 0$$

$$y_5 \geq 0$$

$$y_6, y_7 = \text{unrestricted}$$

The optimal value of a dual variable indicates how much the objective function changes with a unit change in the associated RHS of constraint, provided the current optimal basis remains feasible. The optimal values of dual variables are called "shadow prices." **Shadow or dual price** is the incremental change in optimal objective function for one unit increase in the RHS value of constraint. It is the marginal value of one additional unit of resource. Shadow price remains unchanged within the range of feasibility, even though optimal solution may change [24].

The sensitivity of duals for Run 1 is shown in Appendix 12.E.

It specifies how much the objective function will vary if the constraint value is incremented by one unit. The constraint is active only if the value is nonzero. Therefore, $r2$ and $r7$ constraints are active. If the actual value of $r2$ is increased by 1 unit, then the objective function value will be $(91.4128395 + 1*0.42855 = 91.8413895)$, and if the actual value of $r7$ is increased by 1 unit, then the objective function value will be $(91.4128395 + 1*0.7007 = 92.1198395)$.

Similarly, the sensitivity of duals for Run 2 is shown in Appendix 12.B.

It specifies that the constraint $r2$ is active, as it is nonzero. If the actual value of $r2$ is increased by 1 unit, then the objective function will be $(16.4361664 + 1*4.5454 = 20.9815664)$.

The following points may be noted in the context of primal and dual problems, shadow price, slack variables, and so on [23, 24]:

- When the RHS of constraint represents quantities of scarce resources, the shadow price indicates the unit worth of each resource as predicted on an optimal solution to the primal problem.
- A resource in excess supply is indicated by the **slack variable** for that resource appearing in the final basis at a positive level. In such case, the corresponding shadow price is zero, and additional excess supply is of no value.
- The constraints of the dual problem ensure that at an optimal solution, the profit of an activity can never exceed its true economic worth. Also an activity will never be undertaken if its profit is less than its economic worth.
- The basic relation between the primal and the dual conforms that:
  (i) The minimal cost in the dual problem equals the optimal profit in the primal formulation
  (ii) The vector of shadow prices is the optimal solution of the dual

12.7 LIMITATIONS OF THE MODEL AND SCOPE FOR FURTHER WORK

The present model is essentially based on the classical theory of pricing and is simplistic in its approach. In reality, oil pricing is greatly influenced by the exogenous factors like social, institutional, geopolitics, government policy, market dynamics, and so on. It is difficult to determine these factors and constraints. The model does not take all these exogenous factors into account, although it considers some of these.

The model assumes that price is a linear function of cost, which may not be true in reality. However, LP is good enough for arriving at a satisfying solution for a wide range of problems including the present one. It may be used as a tool to know the preliminary value of base price under different scenarios and can be further fine-tuned for arriving at the desired result and assist in decision making.

LP is an elegant and robust mathematical method, but critics point out the following:

- People in general are not comfortable with such mathematical formulation and find it difficult.
- The linearity assumption of objective function and constraints is often questioned.
- Estimating the constant coefficient values in LP model is not easy as it seems.
- Optimal solution of complex LP model is obtained through trial-and-error approach.

Nevertheless, LP is a powerful technique in solving a broad category of real-world problems.

12.7.1 Scope for Further Work

The model is deliberately kept simple for ease of understanding. Nonetheless, the model is powerful and captures the essential factors of classical pricing theory. Once the essence of formulation is captured, it would not be difficult to expand the model incorporating additional decision variables and constraints.

The *various cost elements elaborated under the paragraph "Cost" in Section 12.4.2 may be incorporated in the model to study the effect of different cost components on oil pricing policy*. Further work can be carried out incorporating noneconomic factors in the model. Developing such a model will be a stimulating experience and will have wider acceptance for functional purpose.

In essence, the model provides a framework to solve similar type of problems for optimizing base oil price and study the effect of varying parameters under different scenarios.

12.8 CONCLUSIONS

Oil is an important source of energy and economic commodity in today's world, which influences the future growth and development plan of a country. Oil pricing is a sensitive issue with far-reaching consequences affecting social, political, and economic spectrum of a country. It is important to have a fair oil pricing policy for balanced and sustainable growth of a nation.

An LP model has been developed in the chapter to optimize base oil price for a country taking into account the cost and share of domestic oil production and that of oil import and other factors. It maximizes profitability and drives oil production by reinvesting profit into E&P activities. It develops a framework to study the effect of various parameters on oil price and aid decision making under changed scenarios and varying circumstances. The model computes base oil price with suitable values of β, P_r, P_m, x_2, x_4, λ, μ, γ, and so on. These parameter values either depend on internal/external factors or are decided by the management based on priority and business need. The base oil price fixed by the state differs from country to country and is not dictated by economic criteria alone; many sociopolitical and industrial issues govern it. The model provides important information on shadow or dual price, slack or surplus variables, reduced or opportunity cost, sensitivity including range of optimality and variability, and so on, and explains physical significance of the results.

Although various methods are available for determining or optimizing base oil price, this exercise employs LP to show its ability to solve such problem with relative ease and convenience. This is a demonstrative model based on sound mathematical logic and algorithm capable of handling multiple variables and constraints. The number of variables and constraints are limited in the current model, which can be expanded, and more variables and constraints may be introduced as per the need and objective. The model will be useful for solving large problems containing numerous variables and constraints that may be of interest in the future. Powerful LP packages are readily available nowadays. With little guidance, complex problems can be solved with relative ease and convenience. All these would help users to take appropriate decision under varying conditions.

Finally, it is reiterated that this is an illustrative model that provides a broad framework for optimizing the base oil price for a country and studying the effect of varying parameters to aid decision making. It is further reminded that oil pricing is a complex issue that is also influenced by noneconomic and socio-political factors.

REVIEW EXERCISES

12.1 What are the various pricing strategies you are aware of? What are the factors that influence the pricing of a product?

12.2 Describe the importance of oil pricing. Why it is different from product pricing?

12.3 What are the methods of oil and gas accounting? How it is different from conventional accounting?

12.4 What approach would you follow to optimize base oil price for a country? What are the factors you would consider? Describe.

12.5 Give some examples where linear programming may be applicable to your areas of work or in your business/industry.

12.6 What are the main properties and assumptions of linear programming model?

12.7   As an Operations Research practitioner, what are the procedures you would suggest to formulate and solve a linear programming problem?

12.8   What is the simplex method? What are its advantages and limitations?

12.9   What is a dual problem? How do you formulate it? What is its significance and relation with primal problem?

12.10 Explain the following terms and their significance:
   (a) Slack or surplus variable
   (b) Shadow or dual price
   (c) Opportunity or reduced cost
   (d) Amortization cost
   (e) Depletion cost
   (f) Unbounded solution
   (g) Postoptimal analysis

APPENDIX 12.A   RESULT FOR RUN 1

Global optimal solution found at iteration:		2
Objective value:		91.41284
Variables	**Value**	**Reduced Cost**
X_1	56.07000	0.000000
X_2	32.95000	0.000000
X_3	200.0000	0.000000
X_4	107.3100	0.000000
Row	**Slack or Surplus**	**Dual Price**
1	91.41284	1.000000
2	0.000000	0.000000
3	0.000000	0.4285500
4	9.000925	0.000000
5	143.9300	0.000000
6	51.24000	0.000000
7	0.000000	0.000000
8	0.000000	0.7007000

12.A.1   (Run 1) Result: Objective

Variables	Result
	91.4128395
x_2	32.95
x_4	107.31
x_1	56.07
x_3	200

12.A.2 (Run 1) Result: Constraints

Constraints	Results
	91.4128395
$r1$	56.07
$r2$	32.95
$r3$	9.00092499999999
$r4$	−143.93
$r5$	−51.24
$r6$	200
$r7$	107.31

APPENDIX 12.B RESULT FOR RUN 2

Global optimal solution found at iteration:		2
Objective value:		16.43617
Variables	**Value**	**Reduced Cost**
X_1	16.68000	0.000000
X_2	3.616000	0.000000
X_3	200.0000	0.000000
X_4	107.3100	0.000000
Row	**Slack or Surplus**	**Dual Price**
1	16.43617	1.000000
2	0.000000	0.000000
3	0.000000	4.545400
4	0.2438336	0.000000
5	183.3200	0.000000
6	90.63000	0.000000
7	0.000000	0.000000
8	0.000000	0.000000

12.B.1 (Run 2) Result: Objective

Variables	Result
	16.4361664
x_2	3.616
x_1	16.68
x_3	200
x_4	107.31

12.B.2   (Run 2) Result: Constraints

Constraints	Results
	16.4361664
$r1$	16.68
$r2$	3.616
$r3$	0.243833599999999
$r4$	−183.32
$r5$	−90.63
$r6$	200
$r7$	107.31

APPENDIX 12.C   (RUN 1) RESULT: SENSITIVITY—OBJECTIVE AND RANGE OF VARIABILITY

Variables	From	Till	From Value	Till Value
Objective	91.4128395	91.4128395	91.4128395	91.4128395
x_2	0	+inf	−inf	0
x_4	−inf	+inf	−inf	0
x_1	−inf	0	−inf	0
x_3	−inf	+inf	−inf	0

APPENDIX 12.D   (RUN 2) RESULT: SENSITIVITY—OBJECTIVE AND RANGE OF VARIABILITY

Variables	From	Till	From Value	Till Value
Objective	16.4361664	16.4361664	16.4361664	16.4361664
x_2	0	+inf	−inf	0
x_1	−inf	0	−inf	0
x_3	−inf	+inf	−inf	0
x_4	−inf	+inf	−inf	0

APPENDIX 12.E   (RUN 1) RESULT: SENSITIVITY—DUALS

Variables	Value	From	Till
Objective	91.4128395	91.4128395	91.4128395
$r1$	0	47.069075	107.31
$r2$	0.42855	7.105427357601E−15	39.2509625481274
$r3$	0	−inf	+inf
$r4$	0	−inf	+inf
$r5$	0	−inf	+inf
$r6$	0	56.07	+inf
$r7$	0.7007	56.07	+inf
x_2	0	−inf	+inf
x_4	0	−inf	+inf
x_1	0	−inf	+inf
x_3	0	−inf	+inf

APPENDIX 12.F   (RUN 2) RESULT: SENSITIVITY—DUALS

Variables	Value	From	Till
Objective	16.4361664	16.4361664	16.4361664
$r1$	0	16.4361664	107.31
$r2$	4.5454	4.44089209850063E–16	3.66964403572843
$r3$	0	–inf	+inf
$r4$	0	–inf	+inf
$r5$	0	–inf	+inf
$r6$	0	16.68	+inf
$r7$	0	16.68	+inf
x_2	0	–inf	+inf
x_1	0	–inf	+inf
x_3	0	–inf	+inf
x_4	0	–inf	+inf

REFERENCES AND USEFUL LINKS

[1] Nagle, T., Hogan, J., and Zale, J., The Strategy and Tactics of Pricing: A Guide to Growing More Profitably, 5th edition, Prentice Hall, Upper Saddle River, New Jersey, 2010.

[2] Shapiro, B.P., The Psychology of Pricing, Harvard Business Review, July–August 1968, p. 22.

[3] Pricing Strategy, available at: www.bized.co.uk/sites/bized/files/docs/pricingstrat.ppt (accessed on February 26, 2016).

[4] Gallun, R.A. and Stevenson, J.W., Fundamentals of Oil and Gas Accounting, Pennwell Publishing Co., Tulsa, OK, 1986.

[5] Wikipedia, Linear Programming, available at: http://en.wikipedia.org/wiki/Linear_programming (accessed on February 26, 2016).

[6] Bazaraa, M.S., Jarvis, J.J., and Sherali, H.D., Linear Programming and Network Flows, 4th edition, John Wiley and Sons, Inc., Hoboken, NJ, 2010.

[7] Bertsimas, D. and Tsitsiklis, J.N., Introduction to Linear Optimization, Athena Scientific, Nashua, NH, 1997.

[8] Chvatal, V., Linear Programming, W.H. Freeman, New York, 1983.

[9] Dantzig, G.B., Linear Programming and Extensions, Princeton University Press, Princeton, NJ, 1963.

[10] Gass, S.I., Linear Programming: Methods and Applications, 5th edition, McGraw-Hill, New York, 1985.

[11] Ignizio, J.P. and Cavalier, T.M., Linear Programming, Prentice Hall, Upper Saddle River, NJ, 1994.

[12] Murtagh, B.A., Advanced Linear Programming: Computation and Practice, McGraw-Hill, New York, 1981.

[13] Murty, K.G., Linear Programming, John Wiley & Sons, Inc., New York, 1983.

[14] Nash, S. and Sofer, A., Linear and Nonlinear Programming, McGraw-Hill, New York, 1996.

[15] Nering, E.D. and Tucker, A.W., Linear Programs and Related Problems, Academic Press, Boston, MA, 1993.

[16] Saigal, R., Linear Programming: A Modern Integrated Analysis, Kluwer Academic Publishers, Dordrecht, 1995.

[17] Schrijver, A., Theory of Linear and Integer Programming, Wiley Interscience, New York, 1986.

[18] Taha, H.A., Operations Research: An Introduction, 9th edition, Prentice Hall, Upper Saddle, NJ, 2008.

[19] Thie, P.R. and Keough, G.E., An Introduction to Linear Programming and Game Theory, 3rd edition, John Wiley and Sons, New York, 1988.

[20] Vanderbei, R.J., Linear Programming: Foundations and Extensions, Kluwer Academic Publishers, Boston, MA, 1996.

[21] Wagner, H.M., Principles of Operations Research, Prentice-Hall of India, New Delhi, 1980.

[22] Naill, R.F., The Discovery Life Cycle of Finite Resources—A Case Study of U.S. Natural Gas, in Towards Global Equilibrium; Collected Papers, Meadows, D.L., eds., MIT Press, Cambridge, MA, 1974, pp. 213–258.

[23] Benjamin V.R. and Kahn, M., Formulation and Analysis of Linear Programs, 2005, available at: http://web.stanford.edu/~ashishg/msande111/notes/chapter4.pdf (accessed on February 26, 2016).

[24] Linear Programming Sensitivity Analysis, Baskent University, Ankara, available at: www.baskent.edu.tr/~sureten/MS(lpsenstivity1).ppt (accessed on February 26, 2016).

[25] Lahaie, S., How to take the Dual of a Linear Program, 2008, available at: www.cs.columbia.edu/coms6998-3/lpprimer.pdf (accessed on February 26, 2016).

INDEX

Optimization and Business Improvement Studies in Upstream Oil and Gas Industry, First Edition.
Sanjib Chowdhury.
© 2016 John Wiley & Sons, Inc. Published 2016 by John Wiley & Sons, Inc.